MW00837695

Climate Change and Globalization in the Arctic

Climate Change and Globalization in the Arctic

An Integrated Approach to Vulnerability Assessment

E. Carina H. Keskitalo

London • Sterling, VA

First published by Earthscan in the UK and USA in 2008

ISBN: 978-1-84407-528-7

Typeset by Composition & Design Services, Belarus
Printed and bound in the UK by TJ International, Padstow
Cover design by Susanne Harris

For a full list of publications please contact:

Earthscan
14a St Cross Street
London, EC1N 8XA, UK
Tel: +44 (0)20 7841 1930
Fax: +44 (0)20 7242 1474
Email: earthinfo@earthscan.co.uk
Web: **www.earthscan.co.uk**

22883 Quicksilver Drive, Sterling, VA 20166-2012, USA

Earthscan publishes in association with the International Institute for Environment and Development

A catalogue record for this book is available from the British Library

Library of Congress Cataloging-in-Publication Data

Keskitalo, E. C. H. (Eva Carina Helena), 1974-
Climate change and globalization in the arctic : an integrated approach to vulnerability assessment / E. Carina H. Keskitalo.
p. cm.
Includes bibliographical references.
ISBN-13: 978-1-84407-528-7 (hardback)
1. Climatic changes--Arctic regions. 2. Globalization--Arctic regions. 3. Arctic regions--Environmental conditions. 4. Environmental protection--Arctic regions. I. Title.
QC981.8.C5K44 2008
551.6911'3--dc22

2008001742

The paper used for this book is FSC-certified and totally chlorine-free. FSC (the Forest Stewardship Council) is an international network to promote responsible management of the world's forests.

Contents

List of Figures and Tables

Figures

Tables

Acknowledgements

Much of the empirical work for this study was undertaken as part of the EU's Fifth Framework Integrated Project BALANCE (project number EVK2-2002-00169). I would like to extend my thanks to Project Coordinator Manfred Lange, Senior Advisor Rik Leemans, and to Monica Tennberg, Nicole Ostländer and my other colleagues in the project. Thanks are also due to Katarina Eckerberg, Camilla Sandström, Jon Moen, Knut Heen, Diana Liverman and Chris West, all of whom patiently commented on earlier versions of the draft, and to the many people who have commented on parts of the work at conferences and elsewhere. All remaining errors are my own.

Introduction

Vulnerability Assessment in the Context of Multiple Levels and Impacts

Understanding change and societies in flux

Change and flux are two of the concepts that best describe the state of society and the environment today. The last generation or so has witnessed extensive structural transformations: communism has fallen, market capitalism and democracy have spread, and economies and corporations have increasingly gone international. States have lost some of their sovereignty over economic and political changes to private actors and international organizations. Many of these transformations have been described under the heading of globalization, which is seen as the intensification, deepening and broadening of international ties and an increased awareness of the world as a whole (Keohane and Nye, 2000). At the same time, environmental change has become increasingly pronounced, as seen, for instance, in the widespread awareness of the pollution caused by industrialization and of its effects such as climate change. Human vulnerability to natural events and changes has been underscored by droughts, floods and changes in weather patterns – problems that could all change in intensity and frequency as climate change progresses. Adaptation to climate change is important, because, among other reasons, large-scale changes are unlikely to be manageable through mitigation (emission reduction) alone; mitigation will not be able to halt climate change in the near future, as this change will be the result of emissions that have already been released into the atmosphere (Smit and Wandel, 2006).[1]

Yet, despite the severity of the changes to come, little research can be found that describes them in the detail required to understand how people conceive of their own situations, vulnerability and possibilities to adapt. Globalization and climate change are often described on a sweeping scale that makes them seem inevitable and abstract rather than concrete and at least to some extent manageable. Changes are also often described in a blanket fashion, with little attention paid to the various ways in which social groups, regions and societies are impacted (Easterling, 1997). Yet the processes of globalization and climate change do not take place in an undifferentiated fashion: they are interpreted and acted upon differently by people who emphasize different economic, political or environmental factors and act upon these in keeping with their local conditions (Eagly and Kulesa, 1997; Tenbrunsel et al, 1997). Changes and trends are thus driven by not only large, abstract mechanisms but also local and regional decision-making processes and economies (Easterling, 1997; Smit and Wandel, 2006).

This work examines change through the eyes of local people, as narratives of perceived vulnerability and the capacity to adapt to change. The term 'vulnerability' is used to refer to a system's susceptibility to change, which has often been seen as the sum of its sensitivity to exposure to change minus its adaptive capacity, that is, its capacity to respond to the impacts of change. Here, vulnerability will be applied as an overarching concept denoting the risk to a community or area; the concept thus underscores social vulnerability and includes climate change as an added specific vulnerability. Adaptive capacity will be seen as the capacity for coping with or responding to change within an area, community or economic sector over the short or longer term. Adaptive capacity is seen as the potential for coping and adjustment that allow for particular adaptations to a number of concurrent stresses. The work develops and applies a qualitative methodology for vulnerability assessment which draws on stakeholders' own experiences to provide a detailed account of climatic changes and of the broader context that determines and constrains adaptation to change. The research also develops a conceptual framework for organizing narratives of change that views local vulnerability in an integrated sense as being limited and determined within a system of multi-level governance.

The study focuses on multiple impacts or exposures through what is known as 'multiple-impact monitoring'; the aim is to describe major relevant changes, how they affect selected regions and sectors, and how these can adapt if the changes continue. It thereby builds on the work in the 'double-exposure' project of assessing how not only climate change but also the concurrent trend of globalization impact actors simultaneously (O'Brien and Leichenko, 2000). The present study starts from stakeholders' perceptions of ongoing change – often describing profound economic and political shifts across the last generation – and assesses these against the changes associated with globalization in the literature.

The work specifically targets interaction across scales: one of the crucial questions is to what extent local adaptation to perceived trends is possible. Vulnerability is often seen as location-specific – as manifesting and requiring adaptation at the local or local/regional scale (Wilbanks, 2002; Naess et al, 2006). Accordingly, vulnerability is often best understood through locally based studies (cf. Easterling, 1997; cf. Dürrenberger et al, 1999; Turner et al, 2003).[2] However, given the progress of globalization, even local systems may be governed to a large extent by international processes, meaning that the local level may be reliant on governance or decision-making processes at levels which may not be accessible to local actors (Smit and Wandel, 2006). For instance, Mariussen and Heen note that decision-making power in large systems has a capacity to 'deconstruct' actors to the point where adaptation to change is no longer possible (Mariussen and Heen, 1999).

As Smit and Wandel note, this means that the question of adaptation hinges on 'the role of local initiatives relative to transformations of geo-political-economic systems' (Smit and Wandel, 2006, p289). This question of scale also applies to the level at which solutions are sought, for instance to problems of resource management in the face of climate change. Many theories regarding resource management confine themselves mainly to lower scale levels. Co-management, for instance, focuses mainly on fair management as a viable process at lower levels; it includes different levels of state management but seldom embraces the possibility that international processes may have a large impact on local management (Sandström and Falleth, 2008). Taking scale into account more closely may round out studies of community adaptation, which have often viewed adaptation as taking place within a single community and thus failed to relate it to the need for feedback between levels and, for instance, national and international decision-making networks. If globalization and, in particular, international actors and systems have a large impact on a local area, this limits the extent to which local resource conflicts (for instance those prompted by climate change) can be solved at that level. In such circumstances, the conflicts have to be defined and resolved in systems of regulation at higher levels. On balance, it is relevant to ask to what extent 'the local [is] the optimal level of adaptation' (Naess et al, 2005, p125). The study will problematize to what extent local adaptation is possible and, if it is, by what means. This includes calling into question the extent to which local adaptation is determined by processes on other levels, including the international, and the extent to which local actors may use these levels *for adaptation or to exert influence in order to enhance their adaptive capacity*. Inter-scalar capabilities, in other words the capacity to act across scales to improve one's own adaptive capacity (Lebel et al, 2005), may be differentiated across populations and even at the local level: some groups may have better-developed networks and thus be better able to act to improve their situation. Such differentiation or fragmentation locally may also impact the way resource conflicts play out locally and change the vulnerability and adaptive capacity of actors.

Research questions

Specifically, the study addresses the following questions:

- What are the current state and the trends of change in the case study areas and sectors, as perceived by local actors, and to what extent, if any, might these reflect broader trends of globalization?
- What increased environmental impacts or continued trends can be expected in general by interviewees or anticipated in light of the climate change impacts and scenario literature?
- In what ways, and to what extent, are local communities vulnerable and able to adapt to these impacts or trends, and to what extent is adaptive capacity

situated at higher scale levels? This question will include assessing the governance networks on multiple levels that affect stakeholders and determine which occupations local communities, or segments of them, may draw upon – and to what extent – in order to increase adaptive capacity.

As the above questions show, this work focuses on how local stakeholders themselves define vulnerabilities, adaptive capacity and adaptations, with stakeholders defined as those groups with an investment and a direct interest in the environment and the use and management of resources within a specific sector and area (cf. Mostert, 2002; Keskitalo, 2004a). The study draws on a multi-level or multi-scale framework in that it engages in local-level case studies and an open enquiry into the changes and agents of change as these are perceived among local people; at the same time, it takes account of regional, national and international levels that have an impact on the case study areas. The study is thus integrated across levels (in taking into account the multiple levels that stakeholders see as influencing them), types of changes (covering the diversity of change described by stakeholders) and types of adaptation (including adaptations and perceived limits to adaptation to these multiple changes, which include but are not limited to climate change). This is accomplished by asking the local respondents to define the actors, authorities, groups and features on different levels that affect them and their daily work. In interviews and stakeholder meetings ('focus groups' in social science terminology), local actors were asked to describe and define the changes they have seen during their working lives. Generally speaking, these accounts describe profound economic and political changes during the last generation. The stakeholders were then asked how they have adapted to these trends and how they can adapt if the trends continue. They were also asked how specific projected climate changes would impact them and how they could adapt to such impacts (their responses often describe perceived ongoing changes and adaptations). The interviewees' descriptions of the decision-making or governance network impacting them – comprising not only government or administrative actors but companies and external organizations – illustrate broad effects resulting from various levels of activity; these include the impacts of international companies and international trade and even the influence of international norms and conventions that shape local and indigenous practices (even those often thought of as practised in a relatively traditional or subsistence-oriented manner) by influencing access and rights to resources.

The case studies in this volume examine forestry, fishing and (mainly indigenous) reindeer herding in northernmost Europe – northern Norway, Sweden and Finland. The selection of these three sectors corresponds to that in the Arctic Climate Impact Assessment (ACIA), a major study of climate change covering the northern areas of eight countries (the Nordic states, the US, Canada and Russia) (ACIA, 2004, 2005). However, given its broad scope, the ACIA deals with the sectors in relatively general terms and its treatment of vulnerability is limited. The present study builds on the work of the ACIA but undertakes a more detailed vulnerability assessment of the sectors identified as important for northern areas. The impacts of globalization and climate change may also be especially clear and easily observable in these areas and sectors. Forestry, fishing and reindeer herding represent parts of relatively small local economies in what are very sparsely populated areas; these are typically more vulnerable than those that enjoy larger-scale economies, which provide more extensive market and employment possibilities. Northern areas will also be impacted early by climate change, the effects of which may be especially marked on the viability of sectors based on natural renewable resources (cf. IPCC, 2001; ACIA, 2005). The specific case study areas in northern Norway, Sweden and Finland have been chosen on the basis of their importance for the selected sectors as well as for the potential they have for illustrating interlinkage between the sectors.

On the whole, given the broad nature of the study of change undertaken here, which embraces both globalization and global environmental change, this work inevitably speaks to different research communities, and can be read in several ways. Through its focus on globalization, multi-level governance and the impact of international norms on the domestic level, it can be situated within the fields of political science and international relations. The focus on vulnerability assessment and understanding the capacity for local adaptation situates the research in the field of human dimensions of global environmental change and the related approaches of vulnerability assessment and integrated assessment. The study constitutes a relatively early study of stakeholder adaptation in the northern European context; the work done to date on northern areas, especially in the area of integrated assessments, has mainly drawn on examples from Canada and Alaska (Cohen, 1996; Cohen, 1997; Ford and Smit, 2004). The findings also illustrate some of the differences between northern Europe and North America, areas often likened to one another in research dealing with the Arctic (cf. ACIA, 2005; AHDR, 2004). Although the book deals with what is in some, mainly political, definitions seen as 'the Arctic', the European areas discussed are at most sub-arctic climatologically. The study thus illustrates the differences between the Arctic areas of, for instance, Canada and Greenland, and those of northern Europe. One crucial contrast, especially with regard to the indigenous, largely subsistence-based economies in Canada, is that the largely indigenous practice of reindeer herding in northern Europe takes place for the most part within a market system. The relatively clear and uncontested differences to be found between indigenous and local communities in Canada are comparatively lacking in northern Europe (cf. Keskitalo, 2004b). In sum, this study adds a European dimension to northern vulnerability research and Arctic studies (cf. ACIA, 2005; AHDR, 2004). Essentially, the book problematizes the ability of communities and related economies to cope with wide-ranging change and describes the demands which such change places on the communities' social organization and political structure.

Organization of the book

Chapter 1 defines the primary concepts in the work – vulnerability and adaptive capacity – and their use in climate change research and vulnerability assessment. Adaptations will here be differentiated in terms of their availability to individual actors (for example individual economic adaptations) or their dependency on other actors in a larger framework such as the political or market context, rather than on the basis of timescale (short- or long-term coping or adaptation). Drawing upon a discussion of the determinants of adaptive capacity, the chapter focuses on governance as a factor crucial to how adaptive capacity is distributed. Processes of economic and political globalization are seen here as potentially extending and determining the characteristics of the governance networks as well as influencing the norms that structure and the values that inform them. The implication is that the character of international processes may determine to a considerable degree the multi-scalar capacities of local actors. These actors may gain support from processes on other levels, depending on the extent to which they agree with and can utilize dominant norms, and access the organizations embodying these, to support their own claims and increase their own adaptive capacity.

Chapter 2 presents the methodology of the study, which centres on interviews with stakeholders but also includes stakeholder meetings (focus groups) and other methods. Underpinning the work is the assessment of social vulnerability with a focus on multi-scalar and multiple-impacts research. The chapter also reviews climate change impacts and scenario literature on projected changes in the focal sectors and areas, and presents the selection criteria for and characteristics of the case study areas in northern Norway, Sweden and Finland.

The next three chapters centre on the economic sectors described above, examining forestry, reindeer herding and fishing respectively. The first of these, Chapter 3, describes forestry in northern Finland and Sweden. It illustrates the relatively globalized character of forestry today, with its extensive reliance on export sales networks, current business conditions and relatively limited planning for, or awareness of, environmental change. It is a sector where a relatively large degree of adaptive capacity and governance is situated on levels beyond the locality or even the state.

Chapter 4 discusses reindeer herding in northern Sweden, Finland and Norway, illustrating how globalization and climate changes impact on relatively resource-poor actors in particular. In all of the sectors, small-scale businesses with limited capital show a relatively high vulnerability to changes in economic, political or environmental conditions. Reindeer herding, a sector where each herder is in practice an independent entrepreneur, illustrates this tendency clearly. Reindeer herding also exemplifies a sector perceived as having a high vulnerability to climate change, as herding needs to adapt to short-term, even day-to-day, changes in weather.

Chapter 5 discusses small-scale fishing in Finnmark County in northernmost Norway. In analysing disputes about the availability of quotas for small-scale fishing, the chapter illustrates particularly well the political factors that are perceived as determining resource access. In the process it reveals – although this can be seen in the other cases as well – the difficulties that local actors see themselves as having in accessing national frameworks and influencing policy in order to institute changes that would support their adaptation.

The final chapter, Chapter 6, discusses the degree to which interviewees consider it possible for themselves and their communities to adapt to the problems perceived on different levels and within the existing multi-level governance framework. The chapter summarizes and compares the factors of change seen in the different sectors and discusses the extent to which local communities have succeeded in adapting to change and their potential to adapt in the future. It also discusses the extent to which governance can be seen as determining or integrating the distribution of the factors that determine adaptive capacity. Overall, the study illustrates that adaptive capacity and the broader determinants of it are to some extent shaped by the governance framework as well as by the norms embedded in that framework that determine the rules of interaction and the distribution of rights and resources to different groups. Vulnerability and adaptation in resource utilization must be conceived of within complex social and legislative environments on several nested levels as well as in the context of the limitations that a number of concurrent stresses may place on adaptation.

1
Structuring the Conceptions of Change

1.1. Vulnerability assessment

This chapter describes vulnerability assessment through an overview of the concepts of vulnerability and adaptive capacity, with special reference to social vulnerability assessment in community-based contexts and 'double-impact' studies, where globalization has recently come to be seen as a major influence on vulnerability. Vulnerability is defined in this work from a perspective of social vulnerability, with climate change seen as an added impact. The study here views adaptive capacity as individual and collective potential and actual actions over both the longer and shorter term in order to encompass the large scope of adaptive and coping responses. The chapter also problematizes the extent to which a vulnerability assessment can be guided by a focus on more general determinants of adaptive capacity. The argument is that the problem of operationally defining vulnerability and adaptive capacity can be addressed by examining the broad decision-making or governance system that to a large extent can be regarded as determining the distribution of the adaptive capacity among different actors. The composition of the governance system is seen as influenced to a considerable degree by large-scale processes such as globalization, and by the norms relevant for larger-scale actors. A vulnerability assessment must thus include identification and analysis of the relevant governance network, how it is composed and structured in response to broader processes, and the adaptive capacities the network affords to different actors or types of actors.

1.1.1. The concepts of vulnerability and adaptive capacity

The concepts of vulnerability and adaptive capacity are especially useful for a broad study of change targeting multiple impacts and their relative weight in the eyes of stakeholders. The two concepts are particularly well suited to assessing human and environmental changes and trends together, and have both interdisciplinary and lay appeal. However, vulnerability assessment, aimed at defining the vulnerability and adaptive capacity in a system (Smit et al, 2000), is to a large extent a methodology under development. It draws on a number of different disciplines and traditions, and focuses to varying extents on different types of vulnerability (from environmental to social), different scale levels (from local to global) and different stresses (from single to multiple). The concepts of vulnerability and adaptive capacity have been developed in, among others, the fields of environmental risk, health and food security research, as well as in the area of impact assessment and integrated assessment (IA) (Polsky et al, 2003; Naess et al, 2006). On the one hand, integrated assessment is an independent activity that aims at interdisciplinary integration – often based on modelling and complex systems theory – of research on complex environmental problems (Rotmans, 2006); on the other, it may be regarded as one stage and contributing strand of vulnerability assessment research (Füssel and Klein, 2006). One and the same research activity may be considered an instance of both integrated assessment and vulnerability assessment. The present work falls into this category: while it focuses in the main on vulnerability assessment, it also contributes to integrated assessment in addressing the relatively little-studied regional scale (cf. Holman et al, 2005).

In the literature on global environmental change, these varying approaches are to some degree being integrated to explain vulnerability to climate change in particular. Climate change presents a crucial challenge in that it will impact on almost every area of the environment and people's relation to it. In the global environmental change literature, vulnerability has been broadly defined as the 'capacity to be wounded'; it is a measure of the sensitivity of systems to exposure to change, minus the capacity of those systems to adapt to change (Kates et al, 1985, p17). This definition draws attention to both an external side of vulnerability (exposure to risk) and an internal side (a system's capacity to cope and adapt when exposed) (Chambers, 1989). Vulnerability is thus crucially related to both the sensitivity of systems and their adaptive capacity, and draws attention to 'what amplifies or mitigates the impact of … change' and 'channels it towards certain groups, institutions and places' (Downing, 1991, p380, quoted in Rayner and Malone, 1998, p240). It also 'emphasizes the degree to which the risks … can be cushioned or ameliorated by adaptive actions that are or can be brought … within the reach of populations at risk' (ibid.). In other words, the concept of vulnerability is attuned to differences in conceptions between groups and the way these notions depend on locally or otherwise constructed discourses and institutional constraints. Self-definition and prioritization of risk become a large part of what renders a community or groups of actors vulnerable: the definition by the actors involved of what they see as risks or threats – and the prioritization of these as more or less immediate – determines what they may act upon.

The concept of adaptive capacity – the possibility to respond to change and undertake certain adaptations in the process – is thus crucial, as it describes the extent to which a system may decrease its vulnerability by learning and applying new economic, social or political approaches to limit risks (cf. Brooks, 2003). Whereas the early literature in the field of climate change focused on specific adaptations, referring to adjustments in ecological-social-economic systems in response to actual or expected climatic stimuli (Smit et al, 2000), the concept of adaptive capacity has more recently come to embrace the broader ability of a system to cope with change-related risks and opportunities (Smit and Skinner, 2002; Smit and Wandel, 2006). As it targets a broad spectrum of strategies and actions, it is no surprise that adaptive capacity has similarities to a number of other concepts: Smit and Wandel list 'adaptability, coping ability, management capacity, stability, robustness, flexibility and resilience' (Smit and Wandel, 2006). For instance, the terms 'resilience' and 'adaptive capacity' are sometimes used interchangeably (Walker et al, 2002), and resilience has been defined as 'the system's capacities to cope or respond' (Turner et al, 2003), which echoes general definitions of adaptive capacity (Smit et al, 2000). Walker et al (2002), on the other hand, define adaptive capacity as a part of resilience, with resilience being a broader term that includes not only a system's learning related to adaptation, but also its abilities to maintain and reorganize itself.

In this work, the concept of adaptive capacity will be applied throughout to define both (short-term) 'coping capacity', which relies more on existing or somewhat modified capacities to act or respond, and (long-term) 'adaptive capacity'. This choice acknowledges that distinguishing between short- and long-term capacities is difficult in all cases and that adaptation is often dynamic and ongoing, with a continuous element of learning or modification based on existing strategies (cf. Berkhout et al, 2006). For instance, 'adaptation may be autonomous and manifest itself through a modification of coping strategies' (Adger et al, 2004b, p68). Thus, '[t]he extent to which ... adaptations occur is likely to depend on how successful existing coping measures prove – if they are successful and no damage is incurred as a result of the hazard, adaptation is likely to be minimal' (ibid.). Similarly, Brooks notes that the term 'adaptation likelihood' could be used rather than 'adaptive capacity' to avoid a discussion of 'where "inherent" capacity ends and external obstacles to adaptation begin' and to better 'encompass ... determinants at different scales' (Brooks, 2003, pp12–13). As Brooks also observes:

> *The issue of scale leads us to think more carefully about our definition of adaptive capacity: will a system with high adaptive capacity automatically adapt? In other words, is adaptive capacity 'self-realizing'? For this to be the case, the definition of adaptive capacity must encompass all the processes that determine whether or not adaptation takes place, and to what extent, including those associated with different scales and systems, representing the environmental, economic and geopolitical context in which the system of interest is embedded.* (Brooks, 2003, pp12–13)

The term 'adaptive capacity' will be used here informed by this understanding (much like lay actors' understanding of 'what they need to do') to cover different temporal scales of adaptation with different levels of existing and extended, modified or newly learned adaptations, and to include multi-scalar capabilities. Adaptive capacity may thus be conceived of as a dynamic, rather than static, space, as adaptations vary with respect to not only exposure to different types of change itself, but also the conditions that influence the sensitivity of systems and the nature of their adjustments (cf. Berkhout et al, 2006). This means that vulnerability assessment needs to account for a large number of 'contextual' factors beyond the impact in focus: contextual factors may actually be as important as the impact proper of, for instance, climate change if concurrent challenges limit available resources or entirely change the conditions for adaptation. Vulnerability assessment also needs to take into account different scale levels that may impact vulnerability and adaptive capacity. All in all, vulnerability assessment is a very complex endeavour indeed.

1.1.2. The history of vulnerability assessment

In recent years, global environmental change in particular has provided vulnerability assessments with new currency and extended scope. At the outset, the focus in this area was the single stress of climate change, but more recent assessments have included other stresses as well, providing a valuable perspective on large-scale change as a whole. For instance, in a survey of adaptation work to date, Smit and Wandel (2006, p285) summarize one of the fundamental findings by noting 'that it is extremely unlikely for any type of adaptive action to be taken in light of climate change alone'. Work by the IPCC, such as the Third Assessment Report, has given more emphasis to adaptation and multiple stresses (cf. Schneider and Sarukhan, 2001). Additionally, '[i]n an important step, IHDP [The International Human Dimensions Programme on Global Environmental Change] has defined vulnerability as a crosscutting priority for the future activities of all its research groups' (Ostrom, 2001, p1). More recently, attention has been drawn to the need to understand how complex governance systems – the organization of decision-making across both public and private actors and at levels from the local to the international – affect and may moderate or ameliorate change. Another salient consideration is viewing human action as integral to natural systems or as integrated into social-ecological systems (Adger, 2006).

This broadened perspective provides unprecedented opportunities to attain a comprehensive understanding of how change at large impacts the human and natural environment and how adaptation strategies may be formed. However, despite a consensus that the overall decision-making situation and its capacities and constraints must be understood in order to understand such change, even today much of climate change vulnerability assessment remains focused on climate as a single stress. As assessments have only rather recently extended their scope from an exclusive focus on climate change, there is a lack of developed methods for undertaking assessments to cope with the breadth inherent in the concept of vulnerability (Smit and Pilifosova, 2001).[1] An example of both of these points –

the recency of the approach and the lack of established methods and a common theoretical framework – is reflected in a 2003 paper where the authors note that '[t]he premise for the workshop (and by extension this paper) was, if the vulnerability perspective is such a promising and innovative research direction, then how does one do it?' (Polsky et al, 2003, p1). Designing vulnerability assessments that include but also progress beyond climate change is thus very much a developing field.

A problem with extending this line of vulnerability assessment to include factors other than climate change is that adaptation to climate change has at times been taken for granted, being characterized as planned, in other words the result of deliberate policy decisions, or autonomous – private – adaptations that are often triggered by market or welfare changes induced by actual or anticipated climate change. Vulnerability assessment to date has thus only to a very limited extent taken into account the fact that adaptations are often responses to economic or political effects rather than to climate changes as such (cf. Smit and Skinner, 2002). Accordingly, one might want to go beyond climate change, as the total adaptive capacity of a system is determined by the resources available in the living situation as a whole (Smit and Skinner, 2002; Smit and Wandel, 2006). While the focus on such broad-based and policy-relevant vulnerability research has increased in the last few years (cf. Schröter et al, 2005), such research cannot accurately assess vulnerability and adaptation as situation- and place-based phenomena without contextual, qualitative studies involving extensive detail and depth. As the number of factors that may be included in a vulnerability analysis in a multiple-impacts study is almost unlimited, the researcher must have a systematic framework that enables him or her to define and systematize descriptions of vulnerability.

In addition, extending climate change vulnerability assessment to such localized and contextual studies entails problems in that climate change vulnerability assessment and integrated assessment have often focused on large-scale, quantitative and scenario-based studies rather than on the qualitative, localized information that is required to address adaptation within a socioeconomic context that comprises people's perceptions. Climate change vulnerability assessments have often used projected changes and biophysical impact studies as the backdrop for defining adaptation options (Kelly and Adger, 2000). Such approaches have frequently treated adaptation in hypothetical terms, often as occurring in response to average climatic conditions and perhaps a generation into the future (Ford and Smit, 2004; Smit and Wandel, 2006). It has thus not been an accepted practice to start a study by assessing the present characteristics of areas and of actors' vulnerability, which a vulnerability assessment focused on actors' experience of adaptation possibilities, constraints and capacity would require (Ford and Smit, 2004). Critical authors note that while starting by assessing the present situation may 'seem to be self-evident ... in fact it differs from the standard impacts/mitigation research paradigm that begins with a consideration of future climate as characterized in climate scenarios' (Burton et al, 2002, p145). This is largely due to an established tradition in global environmental change research of focusing on the future, by, for instance, modelling and proposing scenarios of climate change impact. However, this approach has recently been challenged in the form of assertions such as '[c]urrent vulnerability, determined by past adaptation and the current availability of coping options, provides a baseline from which a system's future vulnerability will evolve' (Brooks, 2003, p10). Similarly, Lim and Burton note that assessment should pay greater attention to recent climate experience, impacts and adaptation and consider both present and potential future vulnerability so that future policy will be based on present-day experience (Lim and Burton, 2001). Any future development will be mediated by a system's adaptive capacity in the present; in other words the path of action and adaptation that communities take in the present narrows their future options, for instance by developing traditions and conceptions of how things are done (cf. Brooks, 2003).[2] Vulnerability at any point in time may thus be seen as a result of adaptations over time, and while it is crucial to view vulnerability in the context of larger trends, a study of the present may allow insights into the complex factors that determine stakeholders' vulnerability and adaptive capacity.

In this vein, Adger et al, for instance, challenge the future-focused study of vulnerability, stating that '[s]tudies of vulnerability may be carried out without a detailed knowledge of how climate will vary over time':

> *Vulnerability assessments do not require detailed climate information generated by models (which is not available for many parts of the world), and they do not require us to wait until the science of climate 'prediction' is more developed. Adaptation policies may therefore be developed despite the uncertainties inherent in the science of climate change – while a detailed knowledge of likely or potential future climate would be desirable, lack of it need not be an impediment to increasing the general resilience of societies to the types of threat that they may be expected to face in the future.*[3] (Adger et al, 2004b, p6)

1.1.3. The extension of climate change vulnerability assessment to social vulnerability

To integrate such perspectives on the importance of present systems, an approach has developed in the climate change field in recent years that centres on social vulnerability and aims to emphasize the 'human dimension', including social, economic and political characteristics (Kelly and Adger, 2000; cf. Adger and Kelly, 2001). In line with such an approach, which will be generally adhered to here, Kelly and Adger define vulnerability more specifically as 'the ability or inability of individuals and social groupings to respond to, in the sense of cope with, recover from or adapt to, any external stress placed on their livelihoods and wellbeing'

(Kelly and Adger, 2000, p328). Social vulnerability focuses on the risk to actors' livelihoods in particular:

> *Social vulnerability is the exposure of groups of people or individuals to stress as a result of the impacts of environmental change. Stress, in the social sense, encompasses disruption of groups' or individuals' livelihoods and forced adaptation to the changing physical environment. Social vulnerability in general encompasses disruption to livelihoods and loss of security.*[4] (Adger, 2000, p348)

This approach thus includes a focus on economic (or subsistence, where relevant) factors: 'the economic capital base of the household, which is essential for the pursuit of any livelihood strategy' (Nyong et al, 2007).[5] This also means that vulnerability may be specific not only to a locality but to a household or even an individual, depending on the conditions in which vulnerabilities occur and the different resources for and extents of adaptation that are available (for example own financial resources, available benefits, available support networks or niches of economic diversification, and markets; Adger et al, 2004b). Sometimes, in order to highlight the risk to individuals' livelihoods, a focus is also placed more on adverse impacts than on the positive prospects change may yield (Kelly and Adger, 2000).

The social vulnerability approach thus focuses on vulnerability as an overarching concept rather than as a residual to be dealt with only when adaptive capacity has been subtracted from the risks. In this case, vulnerability is not dependent on predictions of adaptive behaviour but is primarily ascertainable from the existing structure of the community or other focal entity, the relevant decision-making networks and, accordingly, the economic and political capacity of the entity to respond to risk (cf. Kelly and Adger, 2000). This approach thus emphasizes the comprehensive and inclusive character of vulnerability and rejects the notion that vulnerability can be conceived of in part or only in terms of pre-defined possible adaptations (Kelly and Adger, 2000). Nyong et al (2007) highlight this differentiation from what has sometimes been called biophysical vulnerability:

> *Biophysical vulnerability is concerned with the ultimate impacts of a hazard event, and is often viewed in terms of the amount of damage experienced by a system as a result of an encounter with a hazard. Social vulnerability on the other hand is viewed more as a potential state of human societies that can affect the way they experience natural hazards.* (Nyong et al, 2007; cf. Adger et al, 2004b)

Clearly, it is important in the case of a hazard to view the interaction of social vulnerability with biophysical vulnerability, or the interconnection of socio-ecological systems. Assessments focused on the existing situation may also view the impact of specific stresses.[6] This perspective is reflected in the definition used by Adger et al (2004b):

> *In this report the term 'social vulnerability' is used in a broad sense to describe all the factors that determine the outcome of a hazard event of a given nature and severity (in other words the nature of the hazard is prescribed and the range of possible outcomes of this specific hazard is a function of social vulnerability). Social vulnerability encompasses all those properties of a system independent of the hazard(s) to which it is exposed that mediate the outcome of a hazard event. These may include environmental variables and measures of exposure. For example the vulnerability of a country to a given hazard occurring over its national territory will be a function of the percentage of the population living in the area affected by the hazard, but also of the extent to which individuals and sub-national systems within this area are exposed to its first-order impacts. Exposure and the state of the environment within a system will be socially determined to a large extent. Exposure will depend on where populations choose to (or are forced to) live, and how they construct their communities and livelihoods. Environmental variables will vary in response to human activity, as populations exploit resources and manage the environment for their benefit in the short or long term* (Adger et al, 2004b, pp30–31).

A consensus practice of social vulnerability assessment would start by assessing vulnerability in the present situation and the ways that existing policy and development practice serve to reduce that vulnerability (cf. Burton et al, 2002). In this type of vulnerability assessment, adaptation is seen as including a fundamental social aspect: it involves various individuals and groups (stakeholders within a sector or area) with different, yet sometimes interrelated, points of view. Decisions regarding adaptation are difficult to typify or categorize, for instance as planned or autonomous in regard to climate change, because many adaptations are made continuously, in an ongoing, 'incremental' fashion, in response to multiple stimuli and conditions. Adaptation thus tends to be 'ad hoc, to assume multiple forms, to be in response to multiple stimuli … and to be constrained by economic, technological and socio-economic conditions' (Smit et al, 2000, p214). In order to evaluate and promote the development of adaptations in practice, it is necessary to recognize which actors are involved and what their roles are with respect to adaptation. Adaptation thus fundamentally takes place in the context of economic conditions and institutional and regulatory arrangements, as well as that of the prevailing technology, government policies, financial systems and social norms. This emphasizes the interlinked role of socioeconomic, political and environmental systems in recognition of the fact that:

> *... in some cases, social and economic systems may actually become so thoroughly adapted to political, cultural and economic stimuli that they are effectively decoupled from the natural environments in which they operate. As a consequence they are increasingly vulnerable to climatic extremes regardless of the future climate scenario.* (Smithers and Smit, 1997, p132)

The crucial focus in social vulnerability as part of assessing vulnerability and adaptive capacity overall is thus the assessment of current conditions, trends and, especially, adaptive capacity. These have, however, often been seen as determined by factors such as 'poverty and inequality, marginalization, food entitlements, access to insurance, and housing quality' (Adger et al, 2004a, p30). Nevertheless, it is recognized that the nature of adaptation is to a large extent political and social:[7] it is based on uncoordinated choices and actions of individuals, companies and organizations on multiple levels ranging from the local to the international (Paavola and Adger, 2002). This necessitates the recognition of the network of actors involved in a particular adaptation option, and of how such adaptations relate to broader decision-making processes (Smit and Skinner, 2002). Social vulnerability assessment, as a broad approach, also avoids fixing the question of the focal scale of the assessment, or integration of scales, noting instead that vulnerability may be differentiated even down to the level of the individual (Adger et al, 2004b).

1.1.4. Community-based and double-exposure approaches

Within the social vulnerability approach, initiatives have emerged urging community-based vulnerability assessment (Ford and Smit, 2004). Such assessment focuses on locality-based vulnerability and aims for a high degree of specificity in trying to define who and what are vulnerable, in what way and to what stresses, and how the relevant community can adapt (Ford and Smit, 2004; Smit and Wandel, 2006). Smit and Wandel describe this practically oriented research direction as follows:

> *It tends not to presume the specific variables that represent exposures, sensitivities or aspects of adaptive capacity, but seeks to identify these empirically from the community. It focuses on conditions that are important to the community rather than those assumed by the researcher or for which data are readily available. It employs the experience and knowledge of community members to characterize pertinent conditions, community sensitivities, adaptive strategies and decision-making process related to adaptive capacity or resilience. It identifies and documents the decision-making processes into which adaptations to climate change can be integrated. It is sometimes called a 'bottom–up' approach in contrast to the scenario-based 'top–down' approaches.* (Smit and Wandel, 2006, p285)

This socially focused, localized direction of vulnerability assessment makes it possible to specify some of the complexities inherent in vulnerability and adaptive capacity; in particular it serves to define social vulnerability at the local level, that is, the scale where it manifests itself and at which adaptive capacity is needed in order to respond to changes. Community-based vulnerability assessment illustrates, for instance, that two communities in the same region may be impacted very differently by change depending on their abilities to compensate for it. A small community with high proportions of its population in already vulnerable economies and little access to compensatory resources might find it more difficult to adapt to change than one located in an economically central, well-resourced area. For instance, while an affluent country may generally be able to mediate and adapt to external impacts through its resources and structures, some regions, localities and social groups within that country may not have the requisite resources to adapt and may thus be particularly vulnerable (cf. O'Brien et al, 2004a).

However, in limiting itself to the local level, community-based vulnerability assessment may also exclude or limit a number of impacts from other levels. Some work on community-based vulnerability assessment has addressed the policy relevance of community-based work in which the results of local studies are used to 'mainstream' climate change adaptation by including it in planning and policy development (Smit and Wandel, 2006). Some stress that the local level remains the most important for adaptation (Adger, 2001); others, such as Marsden (in research on a Caribbean case), note that local adaptive capacity is limited by the fact that the main impact on local resource use is external to the community, even global. Market and natural cycles of uncertainty may impact and amplify each other, resulting in an overall impact larger than each sectoral change alone (Marsden, 1997). The problem of scale and how to deal with scale is thus very much a question to be addressed in community-based vulnerability assessment.

An additional path that research has taken to deal with the concerns in vulnerability assessment overall, particularly regarding multiple stresses, has been that of the 'double-exposure' project (O'Brien and Leichenko, 2000; O'Brien et al, 2004b). Its aim is to assess the combined effects of not only climate change but also globalization, since adaptation to climate change will be constrained by adaptation to concurrent challenges such as economic stresses. O'Brien and Leichenko (2000) argue, for instance, that the processes of globalization and climate change – supra-level trends that manifest themselves at the local level – may each create 'losers', in other words groups who are negatively impacted by both and will thus find it hardest to adapt. But they also note that the processes may create 'winners'– people and communities who may find it easier to adapt or may even realize benefits. O'Brien and Leichenko argue that it would be important to identify these losers and winners, as resources for adaptation will be determined by both climate change and globalization simultaneously. The acknowledgement that globalization plays a crucial role for adaptive capacity has also become more widely recognized; for instance Young et al (2006, p304) note that 'globalization is a central feature of coupled human-environment systems' (cf. Smit and Wandel, 2006; Adger, 2006). In more recent work, O'Brien et al (2004b) have developed double-exposure studies in an approach that, first, develops national vulnerability profiles for climate change and for globalization (conceived of as liberalization of agricultural trade) separately; then applies these to identify 'doubly-exposed' districts; and, finally, conducts case studies in selected areas (O'Brien et al, 2004b). O'Brien et al have here measured adaptive capacity in terms of biophysical, socio-economic (human and social capital, access to alternative economic activities) and technological (availability of irrigation and infrastructure) factors (O'Brien et al, 2004b).

1.1.5. Assessing vulnerability and adaptive capacity

Social vulnerability research, including community-based and double-exposure oriented approaches, situate vulnerability assessment within a multi-impact and localized, as well as, to some extent, multi-level framework, making it increasingly complex. While the focus on place-based work has emerged as a consensus in recent vulnerability assessment (Turner et al, 2003; Schröter et al, 2005), with an emphasis that place-based work should be seen in the context of multiple scales, little has developed by way of methods for including the relevant scales (Paavola and Adger, 2002). Determining and assessing vulnerability-limiting or adaptive capacity-enhancing measures thus requires finding a way to deal with and systematize very complex systems in terms of impacts and scale levels, as well as the types of actors involved in adaptation. This has prompted the question of whether the complexity of vulnerability and adaptive capacity preclude their detailed study – a criticism that has followed vulnerability research overall for a long time (Timmerman, 1981; cf. Patt et al, 2005). Patt et al (2005), for instance, note that the task is complex – in particular accessing data on interactions between different drivers of vulnerability – and that the timescale of analysis is too long to be able to make robust predictions about future adaptive capacity. However, others argue that a focus on social vulnerability, while not reducing the complexity of the research, is able to integrate complex data in the form of stakeholders' own understandings on the drivers they face. Regarding the timescale of analysis, Pelling et al (2007) note that rather than, for instance, developing static typologies for adaptive action – 'reactive, concurrent or anticipatory, spontaneous or planned, short-term, and tactical or longer-term and strategic' – researchers should focus on 'the underlying institutional arrangements ... that give shape to adaptive capacity and so prefigure adaptive action' (Pelling et al, 2007, p7). Assessments based on social vulnerability may thus identify institutional impediments to adaptive capacity, such as the need for cross-scale interaction. In addition, the concepts of vulnerability and adaptation may yield benefits when used as overarching concepts that are confined to pre-defined stresses: vulnerability assessment manages, or at least promises, to incorporate the wide range of factors that limit vulnerability, something which no single theory in the social (or natural) sciences is able to do, as such theories rest upon limited disciplinary assumptions that target only one or several of the relevant determinants.

Even if social vulnerability assessment is able to integrate data by focusing on stakeholders' understandings, the complexity of the assessment is daunting. In particular, there is a tension between, on the one hand, the conviction in many studies that vulnerability research requires local specificity and a case-by-case investigation of selected relevant stresses, and, on the other, a desire to determine more general characteristics or determinants of adaptive capacity and vulnerability that can be generalized to theory, can be applied to other cases and can facilitate the assessment of vulnerability. This problem is connected with the complexity of vulnerability assessment, the number of issues it potentially covers and the range of scales that impact vulnerability.

To clearly address the issue of 'vulnerability in regard to what', some researchers have tried to develop frameworks for vulnerability assessment to define the focal system, the attributes of the system that are considered particularly vulnerable, the hazards to which they are vulnerable and the timescale in which the vulnerability is expected to materialize (Füssel, 2007). Other approaches take this even further, stating that vulnerability assessments should 'focus on assessing the susceptibility of specific variables of concern (for example food supply or income), which are believed to characterize the wellbeing of a specific people or place, to a specific damage (for example hunger)' (Luers, 2005, p216). This approach, however, undermines the prospect of a broad and inclusive vulnerability assessment by potentially exempting variables and types of damage that local stakeholders may view as determining in part their adaptive capacity and vulnerability (cf. Naess et al, 2006). Naess et al, for instance, address this shortcoming by asking:

> *How can vulnerability indicator assessments meaningfully represent the local situation?... What outsiders (including climate change researchers) deem most important might not be viewed as most important locally, and finding indicators that reflect local concerns and decision-making processes is a key challenge.* (Naess et al, 2006, p225)

In response to these questions, some scholars have attempted to distinguish between generic determinants of vulnerability and specific determinants 'relevant to a particular context or hazard' (Brooks et al, 2005, p153). The generic determinants of adaptive capacity are often related to general developmental factors and often of a very high generality. It has been recognized that the determinants of adaptive capacity are 'essentially the same' as those for mitigative capacity, in other words the capacity to successfully mitigate climate change by, for instance, limiting pollution (although mitigative measures may be most relevant on higher scale levels; IPCC, 2001), and involve 'similar requirements as the promotion of sustainable development' (Smit and Pilifosova, 2001, p904). This generality could also mean that similar determinants are also relevant in determining winners and losers in the process of globalization. This in turn indicates the extent to which the determinants of adaptive capacity reflect general institutional system capacities, where all the broad types of determinants may have some impact in each case, but where the way in which they manifest themselves may differ between cases (Smit and Wandel, 2006). Additionally, identifiying the determinants of generic adaptive capacity may say little about how the determinants and their interrelations are to be analysed in a specific case.

Obstacles to adaptation or factors that increase vulnerability relate to limitations with respect to the above determinants, such as poverty, lack of knowledge, weak institutions, insufficient resources, infrastructure and incentives, and poor governance (Leary et al, 2007).[8] Similar factors are identified by Tol and Yohe, with the addition of a focus on 'the ability of decision-makers to manage information, the processes by which these decision-makers determine which information is credible and the credibility of the decision-makers themselves' as well as 'the public's perceived attribution of the source of stress and the significance of exposure to its local manifestations' (Tol and Yohe, 2007, p219). Tol and Yohe thus

Table 1.1 *Determinants of adaptive capacity*

Determinant	Encompasses
Human capital	Knowledge (scientific, 'local', technical, political), education levels, health, individual risk perception, labour
Information and technology	Communication networks, freedom of expression, technology transfer and data exchange, innovation capacity, early warning systems, technological relevance
Material resources and infrastructure	Transport, water infrastructure, buildings, sanitation, energy supply and management, environmental quality
Organization and social capital	State–civil society relationship, local coping networks, social mobilization, density of institutional relationships
Political capital	Modes of governance, leadership legitimacy, participation, decentralization, decision and management capacity, sovereignty
Wealth and financial capital	Income and wealth distribution, economic marginalization, accessibility and availability of financial instruments (e.g. insurance, credit), fiscal incentives for risk management
Institutions and entitlements	Informal and formal rules for resource conservation, risk management, regional planning, participation, information dissemination, technological innovation, property rights and risk-sharing mechanisms

Source: Reproduced from Eakin and Lemos (2006), p10, who adapted it from Smit and Pilifosova (2001)[9] and Yohe and Tol (2001)[10]

include values such as perceived credibility. They also note that different determinants of adaptive capacity may compensate for each other. For instance, 'social capital matters most when education levels are low *or* when law enforcement is weak' (Tol and Yohe, 2007, p219). Similarly, social networks may, as emphasized for instance by Pelling et al (2007), play a larger and even compensatory role in societies that are less based on the individual and more on community or kinship networks, or where formal redistributive measures or social security are lacking.

The type of compensation that is effected, as well as which of the characteristics determining adaptive capacity are most important in a specific case, is dependent, among other things, on the type of society (whether a country is an industrial or developing state, for instance). People specialized in a particular economic activity (such as in the industrial world) rather than living in a subsistence economy (such as in areas in the developing world) may have a more limited capacity to adapt to changes that affect their livelihood (Adger, 2000; Turner et al, 2003; cf. Moench and Dixit, 2004). Indicators of adaptive capacity for an industrial state are discussed by Naess et al (2006), and illustrated in the table below. Employment and employment forecasts might not have the same explanatory value in a more subsistence-based society, and the tax base may only be important where a particular social system is in place for the redistribution of wealth. Certain of the more general factors may thus be made more relevant to specific societies.

Adaptive capacity is thus determined largely by the interaction of factors whose importance differs between societies and sectors down to the level of the individual. In order to specify the interaction between these factors, researchers have attempted to prioritize the determinants of adaptive capacity. Tol and Yohe have perhaps gone the furthest, suggesting that 'the adaptive capacity for any system facing a vector of external stresses could be explained by the weakest of its underlying determinants – the so called "weakest link" hypothesis' (Tol and Yohe, 2007, p218). In other, similar attempts, many studies have asserted that the most important determinants of adaptive capacity are institutional multi-level capacity for communication and institutional linkage, as well as an understanding of the patterns of linkage and influence and how they may play out in any given situation. This focus on multi-level interaction (or interplay) is relevant since adaptation in one location is often dependent on adaptation in others and this may improve knowledge dissemination (Young, 2002; cf. Leary et al, 2007; Nagy et al, 2007). For instance, 'virtually all significant environmental regulatory problems

Table 1.2 *Indicators of adaptive capacity*

Social sensitivity	Economic factors	Demographic factors
Employment in the sector	Tax base	Age distribution
	Government budgetary transfers	
	Employment forecasts	

Source: Reproduced from Naess et al (2006, p224)

involve multiple scales at which decisions are required, and that coordination of these decisions is one of the major issues in regulatory design' (McDaniels et al, 2005, p9). The need for integration and communication across scale levels, for instance to tap resources or information beyond the direct reach of the actors involved, may be counted as one of the major contributions of adaptation research to date.

However, one salient consideration beyond the difficulties in determining specific indicators or determinants of adaptive capacity is that adaptation measures may also impede or cancel out each other: 'an action that is successful for one individual, organization or level of government may not be classed as successful by another' (Adger et al, 2005a, p78). By extension, this would mean that while vulnerability indicators such as those mentioned above play a role, some of these, for instance those referring to 'organization and social capital', 'political capital', 'wealth and financial capital', and 'institutions and entitlements', cannot be treated abstractly as factors that can be determined in isolation. Rather, attention must be focused on the systematic interaction and co-construction of these factors, which may be specific to particular systems or parts of a system (such as a sector) and, through their significance in these, may to a large extent determine individual adaptive capacity.

1.2. Adaptation as a political process

Given the need to understand interrelations and the construction of vulnerability and adaptive capacity in a system, there is reason to focus more closely on how to study such systems and specify more clearly the 'forces at work'. Vulnerability and adaptation discussions have long exhibited a rather instrumental and management-oriented view of adaptation in social systems[11] and excluded explicit discussions of power and politics from the process of adaptation, despite acknowledging their importance (Brooks, 2003; Thompson et al, 2006).[12] However, one now sees increasing recognition of the issue of extremely complex socioeconomic multi-level systems where communication and integration may be inefficient, integrated only to a limited extent or distorted by power considerations. Adger et al (2005a), for instance, recognize 'three major lessons from the literature on cross-scale dynamics for implementing adaptation across regulatory and stakeholder landscapes' (Adger et al, 2005a, p80). The first is that 'the issue of adaptation can become a crucible for amplifying existing conflicts over objectives between private and public agents'. Second, 'the institutional interactions in adaptation to climate change at different scales are not some natural pattern dependent on the physical risk'; rather, 'the choice of how an environmental governance problem is handled within a jurisdiction is a reflection of the strengths of the interests and power of the actors who define the problem' (ibid.). And third, adaptation across scales further increases this complexity of power relations and integration. Thus adaptation may reduce the vulnerability of those who are best placed to utilize organization – often those in more influential and less vulnerable situations – rather than that of those who are most marginalized (Adger, 2006). Lebel et al (2005) similarly describe how actors may form coalitions around descriptions or 'stories' of a problem – where they find their interests or values represented in polarized debates. In these cases, actors often frame themselves in terms of particular stories to link to other levels of authority (Lebel et al, 2005).

The need to understand the patterns of interaction in any one system are thus paramount; although it is necessary for efficient interaction, it cannot be assumed that 'the lower scale provides information upwards about the feasibility and acceptability of the broader constraints, as well as local constraints that need to be considered by the broader scale' (McDaniels et al, 2005, p11). While managers at a local power plant may have been able to communicate upwards to higher management in matters affecting the production and profitability of the plant (McDaniels et al, 2005), such communication is more difficult in entire sectors. This may be the case, for example, where the issue is not that of a single actor active on several levels, but of several actors. In the public sector, these actors may not be part of the municipal government and thus not have an opportunity to communicate with higher government levels. In the private sector, correspondingly, staff at the local branch of a company may have limited access to higher management. Larger-scale systems may in such cases (as well as in cases of more direct communication) exercise a strong influence on lower scale levels:

> *The greater power of larger places and higher levels has several underlying reasons, including the dependence of local areas on other places; the greater mobilization capacity of interest groups at higher levels; the heterogeneity of interests and attitudes across local areas; and the dominance of national mass media by higher levels.* (Lebel et al, 2005, para 2)

Communication may thus be limited in attempts at integration across scales, and officials, even if they have a responsibility to the local level, may be difficult to reach, overworked, focused on other tasks or even unsympathetic to specific lines of enquiry (Adger et al, 2005b). The extent to which cross-scale communication 'upwards' may be limited is reflected in the fact that, in some cases, regulators seemingly hold 'a "trump card" of access to central government and higher level rule making bodies' (Adger et al, 2005b, para 34).

It is thus crucial to examine the distribution of power and the selection of influential actors for inclusion in networks of governance if we are to assess possibilities for adaptation. The distribution of political power in this fashion is seldom random; rather, it is often related to other aspects of power, such as economic control over resources and production networks or the status inherent in formal institutions owing to their being well established, having an organization of their own and enjoying a mandate conferred by state legislation, policy or praxis.

Solutions to the cross-scale interaction problem that have been advocated include the use of boundary organizations that broker information across scales, the use of 'scale-dependent comparative advantages' such as functional specialization on different levels, and the use of 'enabling policies' that are constructed to support or at least remove obstacles to actions at lower scales (Cash and Moser, 2000). However, in an imperfect administration, such measures may be difficult to implement because of imperfect communication and interest-based power (Adger et al, 2005a). Additionally, local areas may be fragmented in that several different interests conflict with each other, with each sending different messages to higher levels of decision-making (cf. Lebel et al, 2005). Local communities are thus likely to contain power differentials between different social groups as well as differing interpretations of an issue or needs depending on the group (Goodwin, 1998). Somewhat similarly, Leary et al (2007, p9) note the need for incentives for actors at large to adapt to climate change, as 'those who have a direct self-interest in adaptation may be more astute and quicker to respond'. Mendelsohn defines market adaptation as the 'private choices that individual firms and households make for their own benefit in response to climate change' (Mendelsohn, 2006, p204), which indicates the centrality of economic reasoning that some disciplines assume for actors.

Among other things, these tensions problematize the meaning of 'community' when used with reference to a locally situated group: without research to prove the contrary, shared occupancy in a locality should not be taken to mean communication or shared identity between different members of the 'local community'. This in turn prompts the questions of for whom adaptation and the benefit of adaptation should be assessed in the case of divergent groups and of identifying the interrelation of any particular adaptations with those undertaken by other actors at other levels. While authors have noted that, for local actors, 'empowerment may necessitate acquiring the capacity to work across multiple scales' (Lebel et al, 2005, para 4), such actions may sometimes disempower other actors on the same or other local levels, as well as on other scales. The capacity to work across scales may be acquired through interest organizations (representing diverse interests or united on the basis of region) or through links with similar interests elsewhere.

Institutions are a concept often applied in this context, as institutions and institutional linkages may determine who gets included in or excluded from decision-making networks. Institutions are norms and rules that govern behaviour; they may be informal, such as praxis or tradition, or formal, such as organizations with secretariats and employees (cf. Kelly and Adger, 2000). Networks of decision-making may be used or drawn upon to maximize resources: 'institutions generate and construct their own scales and spaces of engagement to optimize their particular strategic interests' (Tompkins et al, 2002, p1100). Institutions may therefore play a 'key role ... both in reducing vulnerability to climate change but also in exacerbating problems' (Fraser, 2007, p508). In this work, the term 'governance' will be used in preference to 'institutions', as the aim is not to analyse the role of one or a few specific institutions but the overall pattern of formal and informal decision-making – governance – that these constitute. This pattern may play a crucial role in determining the adaptive capacity in a locality or among the actors in it, depending on their placement in networks of decision-making and resource allocation as well as their ability to acquire the capacity to work across scales.

1.3. A focus on multi-level governance for vulnerability assessment

The concept of governance highlights the decision-making framework within which adaptations and actions are delimited or facilitated. The concept is particularly suited to describing the multi-scalar character of vulnerability and adaptive capacity. According to Keohane and Nye, the concept of governance includes 'the emergence and recognition of principles, norms, rules and procedures that both provide standards of acceptable public behavior and are followed sufficiently to produce behavioral regularities' (Keohane and Nye, 2000, p20). Governance in such a definition, which includes the legislation and other regulation that structure people's scope of action, has regularly been seen as the province of government. The literature on globalization and internationalization, however, has defined governance as embracing decision-making by the various actors on different levels – including private actors and corporations, interest groups, NGOs, and government. Accordingly, governance is sometimes also referred to explicitly as 'multi-level governance' (MLG) (Boland, 1999; Hooghe and Marks, 2003). A focus on governance highlights the fact that decision-making is produced in multiple ways by multiple parties within networks comprising both public and private actors and does not in principle privilege any one analytical level. Governance may be a result of state leaders shifting responsibility for certain decision-making to sub-national or supra-national actors or of government being overridden or influenced on certain issues or of its being unable to stem the transfer of authority as private actors develop their own internal regulation or practices in a sector (Boland, 1999). These processes create broader and less easily identifiable (although not necessarily less powerful) networks that range across multiple levels and types of actors.

The development of multi-level governance to such an extent is a relatively novel development in a world which has highlighted the role of state actors in policymaking. In the words of Pattberg, private or market actors have in a short time changed from 'being an intervening variable of the international system to establishing rules that exist mainly outside of it' (Pattberg, 2005, p590). This change can be illustrated with developments in forest governance, which is a domestic issue in that all forest is situated within states. This consideration distinguishes forest governance from that of commons such as the atmosphere or oceans and, among other things, has made it difficult to establish a global forest convention. Nevertheless the forestry sector has seen the emergence of a high degree of private market governance in the form of transnational corporations and associated actors that have established their own labels and certification criteria for environmentally and socially responsible wood production. As a primary production sector, forestry is also subject to internationalization where production and refinement are concerned, which means that decision-making at scales and places far removed from the local communities may dictate production constraints. Forest governance may thus to a large extent be removed from the local level and controlled through a network of actors including states (policy and legislation), major market-based corporations and, as a result, a large number of smaller private actors trying to establish economic niche production (cf. Cashore et al, 2004).

The foregoing example illustrates certain characteristics of governance networks that are important for understanding the way in which governance may be organized. Governance networks may be relatively fluid and may function in a piecemeal fashion, controlled, for instance, by actors involved in production in a specific sector rather than by a single authority. Access to one or several partly connected networks should, however, not be assumed on the basis of their fluid character: even though actors may change, a network may have a 'hard core' of actors with clear, even if informal, requirements for who is included. Moreover, in dispersing influence and decision-making, governance may also lose authority, transparency and legitimacy. This has a particular relevance for local actors, who may not be able to influence or demand accountability of the relevant decision-making structures as readily as they could, for example, by filing a complaint with the government. This poses a considerable problem for the legitimacy, formal accountability and 'opportunities for redress' requirements that are regularly demanded of authority (Keohane and Nye, 2000). Governance by a broader and further removed (for example international) set of actors may thus increase the requirements and constraints imposed on the governance that remains at the local level, exacerbating local–global differences. In the eyes of domestic or local actors, this removal from the everyday sphere may make governance seem unpredictable or driven by special interests (Keohane and Nye, 2000).

The concept of governance thus includes a power perspective in that networks are unevenly distributed, may be perceived as unfair – especially to remote, local, actors with limited influence – and entail a specific cultural normative perspective on issues that may not be mirrored at the local level. Outlining and describing the array of potential impacts at the local level and the actors who administer or govern these is thus an important task in helping us understand the organization and interaction of groups across scales. As mentioned previously, governance networks may be conceived of as organizing around 'stories' of a problem where the actors find their interests represented (cf. Lebel et al, 2005), for instance within the relevant economic, political or issue areas for the focal actor. Such 'stories' often embody or target specific norms that point out specific actions and means as being desirable. The concept of norms has been applied at the international level in particular to explain the basis for the unification of interests – economic, moral or other – which can be seen as socially constructed by actors in their social situation. Norms here bind actors together in a common purpose (Keohane and Nye, 2000) and come to work as a standard of appropriate behaviour for actors with a given identity; in other words they take on a socially compelling character for what an actor should or should not do. Norms are often developed by interest groups (in a broad sense of the term), who promote them, often in international organizations, and cause them over time to become enshrined in conventions, laws or economic incentives such as fines or benefits, or to remain an informal but sometimes equally compelling institution. States may, for instance, follow norms (to different extents) in order to gain legitimacy or avoid ridicule (such as diplomatic support or criticism) (Finnemore and Sikkink, 1998; Arts, 2003).

By this process, a norm may eventually become so widely accepted that states as a matter of course take up new responsibilities or endow individuals with new rights according to the norm (Finnemore and Sikkink, 1998). By enforcing norms in the interpretations, or 'stories', that urge one way of acting or attribution of rights in preference to another, actors can change the legitimate arena and issues for decision-making. Domestic actors can use this function against international actors or each other by, for example, making claims and engaging in domestic campaigns that use the norm, by working outside their domestic societies to gain support for their claims at home, or by making a conflict international through cooperation with actors from other countries against the actions of one or another state or international institution (Tarrow, 1999). Norms developed at the international level may also impact interests at different levels, affect economic interests, create fragmentation or unification locally, and empower certain actors or interests to work across scales (cf. Lebel et al, 2005).

As a result, we might expect to see actors even at a local level act in relation to or be governed by a multi-level framework made up of international and national private and public actors. The actors will be organized into networks through economic interests and organizational praxis, including, for instance, different levels of the state, some setting and some implementing relevant regulation, and will be united around stories or interpretations that forward these interests and express different norms for behaviour. Fundamentally, such a governance framework and the norms embedded in it may determine the rules of interaction as well as the distribution of rights and resources to different groups, thereby determining and delimiting a large part of the adaptive capacity available to different actors.

In effect, the governance framework delimits a range of adaptation options, constrained by the limitations on adaptive capacity at any given time, that may be termed the 'adaptation space' (Berkhout et al, 2006). Delineating this space is one of the crucial tasks of a vulnerability study, for it reveals what is unexpected or beyond a system's range or scope of action in coping or adaptation. As Yin et al have observed, 'a system will not be seriously damaged as long as the … change or variation falls within a system's coping range, while changes beyond the boundaries of the coping range can result in severe or even disastrous consequences' (Yin et al, 2007, p89; cf. Smit et al, 2000).

Given a focus on the role of the governance network, division of adaptations by type may distinguish those which actors can undertake and decide on themselves and those that require larger-scale interaction. While individual and larger-scale adaptations may at times correlate with short- and long-term adaptations, they also encompass, to some extent, adaptation likelihood (Brooks, 2003) and dependency as seen by individual actors. One can thus differentiate the adaptation measures that are the responsibility of governance on a broader scale, such as political regulations, and accordingly delimiting individual adaptation. Within the space made available for adaptation by political decision-making and the market, individual actors will most likely implement adaptations on the individual or household level and invoke strategies or modifications of strategies geared to economic or market-based adaptation that may be commonplace to economic actors. The implementation of larger-scale adaptive actions may thus be more uncertain or even unlikely for individual actors unless they are represented by larger-scale interests invested in and able to influence policy or other changes to support the actors. This work will distinguish between individually possible – often individual-level economic adaptations – and systemic adaptation measures that change the 'adaptation space' for individual actors in order to highlight the impacts of the governance system on individual and local adaptation. The research assumes that the ability to adapt and influence systemic constraints to adaptation is unequally distributed between actors, with the adaptation space available to actors on lower scale levels being circumscribed by the governance framework. The extent to which actors on lower scale levels are able to influence the broader governance framework is largely a result of their resonance with and access to norms and the organizations espousing these on higher scale levels. Such access empowers local actors while potentially disempowering others, resulting in differential adaptive capacities and vulnerability sometimes being 'transferred' between actors.

1.4. Analysing, organizing and understanding large-scale change

To assume that even a place-based study has to take account of larger-scale and sometimes international governance is to also assume that global-level processes play a role, or manifest themselves, locally – something that has been noted for climate change and governance on higher levels, as well as for globalization (cf. Rankin, 2003). Globalization may actually play a part in determining the network of governance, as it makes actors at other than the local level relevant and draws attention to international and market actors and processes. Processes of globalization may thereby contribute to delimiting the adaptive capacity available to different actors.

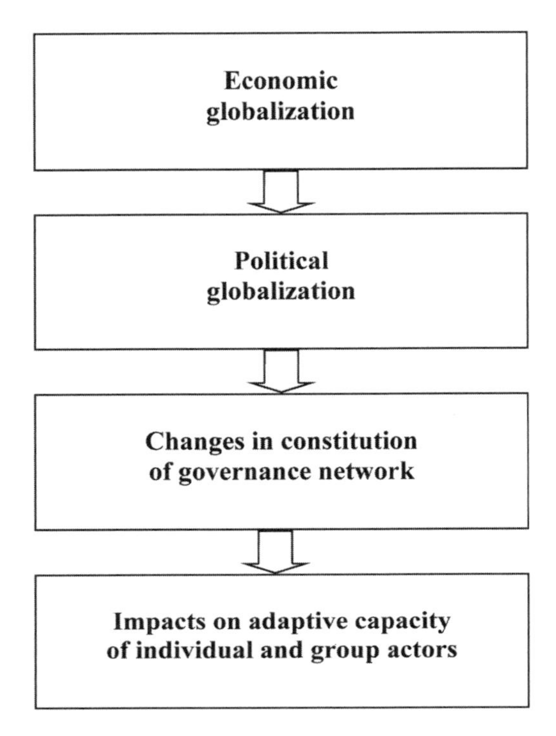

Figure 1.1 *Simplified diagram depicting multiple-level impacts on adaptive capacity*

Effects of globalization may thus come to change the constitution of the governance network, as illustrated in Figure 1.1. This will lead to differential impacts on adaptive capacity of individual and group actors. This is because the concept of globalization, like climate change, does not imply a homogeneity or equity; globalization is differential and may increase rather than decrease social differences such as mobility and access (Keohane and Nye, 2000). In other words, it may increase mobility and interconnectedness for some while increasing the gap between those who are interconnected and those who are not,[13] creating winners and losers in the process and possibly exacerbating impacts in the case of actors influenced by climate change. Globalization is also a multi-scalar issue and can be located on a continuum of different impacts manifesting themselves at the local, regional and national levels (Held et al, 1999). The global scale may here be seen as defined to some extent by a new international division of labour. On the national scale, economic policies can be seen as being undermined by the increasing mobility of companies and global capital. On a regional scale, industrial core areas have experienced dramatic deindustrialization. Regional industrial sites throughout the world economy compete more directly with each other (Brenner, 1999). Localities, too, may be forced by globalization to become competitive on the world market. Even small localities may interact internationally with a multiplicity of different groups of actors, including private ones (Sites, 2000).

Such a situation – to some extent a result of reduced national support for localities, which forces them to become competitive – may cause localities to become less integrated with the regional or national economy and prompt them to operate on a world market instead, sometimes increasing the adaptive capacity of actors, groups or localities able to do so. If they are unable to achieve such integration, on the other hand, they may become marginalized economically and possibly also politically (cf. Pedras, 2002).

Studies of globalization and multi-level governance and of the differential vulnerability such processes result in must thus encompass an understanding of how macroeconomic and transnational processes interact with historically specific practices and ways of life at the local scale. However, the substantive analyses of how these overarching trends affect – if in fact they do – 'ordinary people' and their everyday lives and identities have so far largely been lacking in globalization research. Kennedy, for instance, notes that '[t]he difficulty ... is to gauge how much the mass of ordinary individuals in everyday life across the world perceive, understand and directly experience all of this and what inferences, if any, they may draw as a result' (Kennedy, 2007, p270). Viewing stakeholders' perceptions of change is thus especially relevant in order to assess to what degree their perceptions resonate with the changes associated with globalization; it may also be possible to assess whether and to what extent stakeholders are turning to 'new' frameworks of identity, possibly in order to forward interests that are not supported locally, and thereby enabling multi-scalar interaction capacities. Such a cultural globalization or extension of identity is discussed in much globalization literature, but seldom problematized as to the extent to which it may actually be taking place, or as a result of which processes with which groups or localities (cf. Kennedy, 2007). This is a gap which vulnerability assessment with locally based components could address.

To include such an assessment of globalization in vulnerability assessment, characteristics (or determinants) of globalization would need to be specified. However, the problem of determining 'exposure to globalization' is to some extent akin to that of determining the characteristics of vulnerability at large. Globalization is a complex concept with various definitions, one that can be broken down and analysed in various ways depending on the purpose of the definition used. Some authors note that globalization is the present concept of choice for describing major and poorly understood processes perceived on several sites; in this respect, it has come to replace previous terms, such as modernization, in referring to large-scale societal transitions (Guillén, 2001; Roudometof, 2003).[14] Globalization, however, to some extent highlights processes that are qualitatively different from those that fall under modernization. Most definitions of globalization centre on its tendency to link otherwise distant groups and areas closer together and to make us realize that we are affected by actions far away. For example, Keohane and Nye define globalization in a general sense as the 'increase in a state of the world involving networks of interdependence at multi-continental distances' (Keohane and Nye,

2000, p2; cf. Robertson, 1992; Held and McGrew, 2002). This definition views the networks of interdependence as intersecting more profoundly and at increasingly diverse points today than previously – something that would directly impact the way in which governance networks are constituted. As we can see the beginnings of globalization in Columbus' discovery of America or in the invention of the printing press, what distinguishes globalization in recent times is not that it is a new process but that it is proceeding with increased magnitude, complexity and speed. An example of the political change is that international organizations have multiplied in number and importance, particularly since World War II, to the point where today they are an indispensable part of the international decision-making network (Keohane and Nye, 2000; Held and McGrew, 2002).

As it is difficult to distinguish whether specific changes in a local context are a result of globalization rather than of other trends, domestic structural changes, or restructuring processes or events at other levels (Roudometof, 2003), no direct causal link can be implied between processes observed locally and globally. The cause–effect relationships may be both indirect and unclear: national changes that impact the local level may in turn be caused by globalization but in the particular case could be a result of other processes. A comparison can, however, be made between the problems outlined locally and those outlined in the globalization literature, where similar types of changes across empirical studies of areas and sectors may indicate global drivers. The sections below will define the salient features of globalization for present purposes (a study focused on local renewable resource-based sectors), along with the features of globalization that are generally seen as driving globalization as a whole (economic globalization) and are relevant to a governance perspective in that they add a description of political globalization to that of economic globalization. Both social and cultural features can be understood as underlying the larger economic and political narratives of change in the sectors and as potentially leading in time to a more extensive integration of identity for certain groups.

1.5. Conceiving of multiple impacts: Definitions of economic and political globalization

1.5.1. Economic globalization

For countries where international trade represents a significant proportion of the economic activity, the domestic economy is to a great extent determined by the conditions that obtain abroad (Norén, 2004). Not surprisingly, the economic dimension of globalization has received the most extensive coverage, both within and outside academia (Rankin, 2003), and is often seen as driving many other forms of globalization. 'Economic globalization is generally used as a shorthand term to describe the wide-ranging changes in the world economy and the restructuring of national political economies over the last 25 or so years' (Conley, 2002, p449). Globalization will be used here to refer to both a formal institutional potential for and actual levels of engagement with other economies (Scruggs and Lange, 2002), in other words a process of change from a national to a global scale of integration of production and trade (Sykora, 1994). It is often perceived as a transformation that has become possible through the 'spread of capitalist relations and impersonal market transactions across the world' (Kennedy, 2007, p270), even to the extent that globalization is seen as the latest stage of capitalism, or as the extension of capitalism to the global scale (Kennedy, 2007).

Economic globalization may be seen in, among other things, increased international trade, direct foreign investment, the globalization of financial flows, markets, corporations and consumer patterns, and the diffusion of technology and related knowledge worldwide (Budd, 1998; Guillén, 2001; Mahler, 2004):

> *Economic globalization means that capital has reorganized production on a worldwide basis and in doing so has moved far beyond the previous situation where national capitals established world links but these worked mainly through the trade in finished goods and capital flows based primarily on the search for raw material and colonies. Instead, and for the first time in history, production itself has become globalized, thereby breaking down once national circuits of production and integrating them across borders and nations.* (Kennedy, 2007, p269)

This economic transformation diminishes the role of national governments by taking over some of their regulatory capacities, for instance removing some of the governments' possibilities to regulate their own economies and thereby employment and tax revenues (Budd, 1998). The extent to which these changes can be seen as indicative of *global*ization is disputed, however; rather than being truly global developments, they often indicate a regional *international*ization that, despite its growth, is limited relative to the state. Thus, for example, the volume of international trade in most economies is small relative to the overall size of the economy; domestic investment is greater than foreign investment; and multinational corporations locate most of their assets, owners and main managers in their home countries. This being the case, large areas of the world – such as those lacking these kinds of networks and integration – have not been affected by globalization (Guillén, 2001).

However, for present purposes – a study that centres on local renewable resource-based production – the salient features of economic globalization can be seen as those manifested in international trade networks. Here, multinational corporations (MNCs) are the main example of trade and production globalization. MNCs are regularly seen as companies present in more than their country of origin; the more recent term transnational corporation (TNC) suggests that these companies may be present in a large number of countries, even removing themselves from their country of origin. However, advances in communications technology and infrastructural conditions have contributed to an internationalization of production and marketing among small- and medium-sized enterprises (SMEs) as well, at least in advanced economies (Held et al, 1999). The companies – large or small – participate in or develop 'commodity chains' – networks of labour and production processes whose end result is a finished commodity (Snyder, 1999) and which may be seen as networks of economic interests for the purpose of describing the development of economically oriented parts of governance networks. Commodity chains may include sourcing and production in different areas, sometimes within international production networks having fragmented ownership, as well as the export of the final refined commodity, possibly to become included in new production networks, for example a wood panel becoming part of prefabricated house production. These commodity chains may be producer-driven – controlled by large integrated industrial enterprises – or buyer-driven, where production networks are typically decentralized and the power rests with large retailers (Snyder, 1999).

In an internationalized economy, firms may focus on exporting their products in order to take advantage of the larger export market, on organizing production internationally to take advantage of lower costs abroad for particular stages of the production process, or even on moving operations abroad to circumvent trade restrictions and reduce transport costs (Held et al, 1999). These factors highlight the international character of production and marketing as well as the vulnerability this entails for local areas. For example, as outbound investment removes capital and causes unemployment in the local economy, multinational firms may threaten to relocate in negotiating with their workers or with government in an attempt to drive down wages or tax levels. Moreover, private businesses, even those not directly affected by such processes, may keep workers' pay or benefits down in an effort to maintain competitive prices or raise profits. The risk of companies moving abroad may also limit the possibility of domestic political leaders using macroeconomic policy mechanisms to limit unemployment and transfer social security costs to the individual, thus negatively affecting vulnerable low-income groups (Held et al, 1999; Mahler, 2004). In sum, economic globalization entails external financial impacts on governments that have contributed to both the emergence of a more market-friendly state and a shift in the balance of power between states and financial markets (Held et al, 1999). Globalization may also result in rationalization due to the workings of international competition and dynamics – in which technological change has played a large part – and in an increased distance between employees and economic or company decision-makers.

Economic globalization may thus have distributional effects, impacting the state, its economic policy, its provision of social security and employment levels. It may also affect or prompt processes of structural adjustment among producers (Norén, 2004). Accordingly, the internationalization of production/trade networks in both MNCs and SMEs are among the crucial factors to be assessed in the case of economic globalization. Any assessment must also look at how these international networks and demands lead to a shift in how relevant the location of production is or the degree to which trade networks are detached from the local level. Additional focuses are the changes – if any – that occur in social and economic effects where place-based employment is concerned and the possibilities for adaptation that exist within the economic and socio-political structure. In this work, these questions will be analysed with reference to interviewees' descriptions of changes in their work over time, which reflect international economic influences on their livelihoods, local self-sufficiency or reliance on local resources, and the overall level of change.

1.5.2. Political globalization

Economic globalization can be seen as driving many of the current changes in the world, which states and, to some extent, localities, both play an active role in and have to respond to. Political globalization can be regarded here as a stretching and deepening of global political processes, often in response to economic integration (Held et al, 1999). Political globalization can also be seen as involving the intermingling of layers of power and interest at and across the national and international levels (Shelley, 2000). These developments mean that decision-making actors face increasingly complex situations: the decision-making power of the state at local, regional and national levels may be influenced by decision-making in private and international organizations and institutions, such as the EU, as well as by interest groups affecting public opinion, thereby creating a decision-making field with multiple actors.[15] Together, the driving forces of economic globalization and political globalization could greatly influence the actors and processes that make up a large part of the governance network.

For instance, even though local decision-making and adaptive capacity may to a large degree be determined by state policy and legislative frameworks, some restructuring of the state in response to globalization can already be seen. Certain central state powers have been devolved to sub-state regional and local units, which may then be expanded in geographical scope to achieve efficiency and savings in administration. For instance, some regulation may take place at the regional level to unburden the state; municipalities may be integrated into larger municipal units with more capacities in order to trim costly administration; and local services may be centralized into larger 'local' units in the process. This restructuring can to some extent be seen as a response to states' decreasing economic power, one consequence of which is that sub-state units have not received economic capabilities commensurate with their increased regulatory responsibilities. While 'the current wave of state re-scaling can be interpreted as a strategy of political restructuring that aims to enhance the locationally specific productive forces of each level of state territorial organization' (Brenner, 1999), devolution has not necessarily brought adaptive capabilities to localities with low populations and thus a limited tax base. Instead, economic globalization and the decreased economic means of the state have resulted in 'increased demands on the state to provide social insurance while reducing the ability of the state to perform that role effectively' (Rodrik, 1997, p53, quoted in Mahler, 2004, p1027). As a result, the adaptive capacity of local communities or sectors that rely to a large extent on state support or on workplaces in the public sector for employment may have diminished.

Economic changes thus have political consequences. Government is often seen as having a tripartite role: stabilization through macroeconomic policy, allocation through the provision of social goods, and the redistribution of resources through welfare organization (Budd, 1998). With the first of these roles being eroded, government at any level can only act as a conduit: it can create the organizational basis or infrastructure for local competitiveness but not supply or support the economic means to facilitate a competitive local economy (Budd, 1998). This fundamentally hollows out the role of the state. On the other hand, some liberal authors argue that it is not globalization that is to blame for these changes: that globalization serves as a powerful engine of economic growth and sustained employment while the determination of who become the losers and winners in globalization is a result of domestic political factors such as the political profile of governments and the nature of labour relations (Mahler, 2004). In either understanding of globalization, however, the role of the state has been transformed and to some extent constrained.

In any event, it remains crucial to understand the role of domestic politics – through which many of the impacts of globalization are mediated, although not necessarily tempered – in order to comprehend the impact of change on a specific locality. In either a positive or a negative understanding of globalization, one might say that the present processes include new forms of re-regulation through which states actively promote the present process and features of globalization. The changes include, at the state level, the development of new forms of industrial, technology and regional policy, the construction of new legal and financial regulations, and the establishment of new means to enhance production within selected areas in each state. What emerges, some argue, is a new type of 'competition state', whose central priority is to create a favourable investment climate for transnational capital; consequently, 'the state itself becomes an agent for the commodification of the collective, situated in a wider, market-dominated playing field' (Cerny, 1995, p620, quoted in Brenner, 1999, p65; cf. Cerny, 1997). At the same time, companies and interest groups adapt by organizing internationally for competition or by allying themselves with international interest groups in order to defend local or special interests.

These factors of change and adaptation at multiple levels are relevant in the present study, whose point of departure is the local level. The competition between areas takes the form of competition between economic specializations and operations specific to each locality. Social capital, in other words informal networks of interaction and trust that may be embedded in production and service facilities, integration with international institutions (for instance representing major norms), or major cultural practices, make up some of the basis on which a locality can compete for the location and maintenance of economic activities locally (Budd, 1998). However, small towns are not in a position to take advantage of the dynamic of internationalization in the same way as core cities can; regional hinterlands may struggle with the costs of the decline of traditional industries. 'The lack of a regional stabilization role limits regional and urban policy to a series of defensive responses to supranational and local changes' (Budd, 1998, p682). This is, again, a potential limitation on the adaptive capacity that can be exercised at the local and regional levels and a constraint that indicates the extent to which supra-level trends may be driving local changes.

Of particular interest in a locally based study is the degree to which actors outside the locality have influenced political decision-making, the distribution of power, and employment in the community. It is also relevant to assess the existing network and organization of local and other actors through which any adaptation may be implemented. As stated previously, the focal questions include whether there exists a local network providing connections to organizations in other communities and whether interest groups are well organized and connected to interest groups on other levels, whereby they would attain the capacity to work across multiple levels (cf. Lebel et al, 2005).

1.6. A framework for assessing vulnerability and adaptive capacity to multiple stresses

Overall, the framework developed in this chapter will be utilized to define, first, the impact of economic and political globalization on changes in the areas studied (or at least the co-variance of the process with such changes). It has been argued here that the very composition of the governance network is likely to be affected by globalization in that a number of remote – perhaps even international – actors and elements (such as regulation or norms) will impact local actors. In particular, norms that are important on other scale levels to other actors, and useful for some local actors depending on their livelihood, can be expected to have an impact; this may divide local communities, empowering some stakeholders but not others.

The governance network will thus constitute part of the context that determines actors' vulnerability and adaptive capacity to change, for instance in the form of access to other niches of production, buyers, commodity chains, markets or diversification, or access to elected decision-makers on higher levels. Access to decision-makers could be a possibility for larger actors or larger interest organizations (possibly united on the basis of norms for areas beyond the local) and could, for instance, yield opportunities to amend legislation and the systems of benefits, support or financing in which rights are enshrined to benefit the actors in question. Access to decision-makers, cross-scale communication or integration would, in light of the discussion above, presumably be limited, especially for marginalized groups or groups without access to interest organizations or broader norms through which they can forward their interests. In such cases, actors will be limited to adaptations that are within the means of the individual or the household adaptive capacity and the limited resources of the present system. These resources may be depleted in the case of more severe impacts (for example those of climate change, globalization, or major changes in the governance systems that are intended to benefit other actors or interests). Actors with more limited resources may also have these resources depleted by changes that are persistent over the longer term yet do not prompt actions at higher levels that would raise local adaptive capacity. Access to resources for adaptation is thus fundamentally framed as a multi-scalar issue.

2
A Methodology for Vulnerability Assessment

2.1. The approach to vulnerability assessment in this work

The focus in this study is on governance as the framework that to a large extent delimits adaptive capacity available at the local level. This perspective will be investigated through local stakeholders' definitions of the actors that have an impact on them and their livelihoods. Accordingly, the study places a fundamentally qualitative, social scientific focus on the primacy of stakeholders' understandings. It also highlights and utilizes some of the methodological features of vulnerability assessment discussed in the previous chapter. These include a focus on social vulnerability based on the present and including a specific assessment of particular stresses, in the present case climate change. In addition, the study assesses the impact of globalization, which is seen as affecting the distribution and characteristics of the governance network. In keeping with some consensus in the vulnerability assessment field (Turner et al, 2003; Schröter et al, 2005), the research recognizes the value of a place-based study that includes higher scales, focuses on multiple stresses and notes the differential vulnerability among different types and scales of actors. The study thus includes an approach based on a bottom–up, practical focus in community-based vulnerability assessment but extends this through a multi-level perspective and, to some extent, the inclusion of varying vulnerabilities at the level of individuals or their livelihoods.

As a result, the study aims for integration across levels, types of changes and types of adaptation: the levels take into account the multiple levels that stakeholders see as influencing them; the changes cover the diversity of change described by stakeholders; and the types of adaptation comprise adaptations and perceived limits to adaptation to multiple changes, which include but are not limited to climate change. The work yields knowledge about the multiple scales of decision-making that affect local/regional units by examining interactions and perceived influences (cf. Naess et al, 2005). The research examines clearly defined systems (sectors within a region), recognizes livelihood as a valued attribute, and defines hazards to include the changes and challenges identified by the stakeholders as possible impacts on the community (cf. Füssel, 2007). However, the study does not attempt to pre-define specific variables of concern that may characterize wellbeing in the communities (Luers, 2005), or to hypothesize what stresses stakeholders are subject to (Schröter et al, 2005). Rather, the research allows stakeholders to define the aspects of their situation which they consider vulnerable, to identify the trends that they are vulnerable to and to assess their vulnerability to climate change. Here, as in Smit and Wandel (2006, p289):

> *The goal ... is not to produce a scoring or rating of a particular community's current or future vulnerability. Rather, the aim is to attain information on the nature of vulnerability and its components and determinants, in order to identify ways in which the adaptive capacity can be increased.*

Moreover, the study does not aim to develop indicators for vulnerability or its components (Schröter et al, 2005), although in the final chapter it does discuss the implications of the findings for the general indicators of adaptive capacity described previously. Overall, this relatively open-ended design is intended to make possible a stakeholder-based assessment of vulnerability, of the communities' socio-economically and politically delimited adaptive capacity, of the adaptations that are being or might be undertaken, and of where climate change impacts figure in people's priorities. These components of the study will be discussed below.

2.1.1. Stakeholder-defined nested scale levels and stressors

Current developments in vulnerability assessment have essentially shifted the focus from response to a specific exposure to the broader social vulnerability that in part determines such responses. Attention is also drawn to the role of adaptive capacity and to present situations, as strengthening general adaptive capacity in the present would enhance society's capability to respond to multiple stresses. The emphasis here will be on such a *social vulnerability*, in the sense of the state or present situation that stakeholders identify, and with a specific reference to livelihood, which is analysed in terms of the salient economic sectors in specific localities. Given that there does not exist a well-established approach enabling social vulnerability assessment to deal with issues of multiple or nested scale levels, this study draws upon some of the bottom–up, qualitatively oriented research and stakeholder-focused approach of community-based vulnerability assessment.

The study extends a community-based vulnerability assessment approach by taking into account the complex existing priorities both at and *beyond the local scale*, in other words the forums where many of the locally proposed adaptations would need to be acted upon. This is accomplished by considering the entire decision-making network described by interviewees – the local, regional, national and international levels in the case of multi-level governance – and by examining how higher scale levels determine which actions, in the eyes of the stakeholders, are made possible at the local level. In this regard, the study adds a political, multi-level context to more entitlement-based explanations that often look at class, social status, gender and like factors as determinants of vulnerability (cf. Adger, 2006).

Adger et al (2004b) suggest that a social vulnerability assessment (in their case on the national scale) should be undertaken by first assessing generic vulnerability and adaptive capacity in order to identify ways to cope with 'a wide range of hazards' (Adger et al, 2004b, p39), as well as principal existing hazards or trends that may cause problems in the future. In this context, the stressor of climate change is seen as a potential, and in some cases existing, hazard that represents a 'specific' vulnerability in relation to social vulnerability overall (Adger et al, 2004b). This approach is largely congruent with that applied in this work, where, in broad terms, the present situation is described with a focus on the factors or perceived trends of change that stakeholders see as impacting their vulnerability, or what may be called 'exposures and sensitivities' (cf. Smit and Wandel, 2006). 'Exposure and sensitivities' are seen by Smit and Wandel as 'almost inseparable properties' (Smit and Wandel, 2006, p286) that reflect a system's exposure to a risk or change and its sensitivity to such exposure. In this work, stakeholders are asked to identify both changes (exposure) and sensitivities to these (how such changes affect them). Moreover, stakeholders' descriptions of these changes over time are systematized in terms of the economic changes over time and the political or regulative situation and changes in this, in order to enable comparison with factors of economic and political globalization. The study thus incorporates and extends *double-impact* research (cf. O'Brien and Leichenko, 2000) by including multiple stresses as defined by stakeholders; it considers these in light of the economic and political changes they describe and assesses them for their resonance with changes described under concepts of economic and political globalization.

Stakeholders are also asked to describe their adaptations to the changes over time that they have identified, in terms of the scale levels and actors that affect their adaptation. The description of scale levels and actors is treated as a functional definition of the governance network that stakeholders see as affecting them and against which they position themselves; the network may be reflected as specific norms expressed by the actors in it, such as norms on environmental protection and indigenous peoples' rights. Furthermore, to account for the specific vulnerability related to climate change (Adger et al, 2004b), stakeholders were asked to describe, in response to lay-language formulations of climate change and in light of their own experiences, potential (and sometimes actual) exposures and sensitivities, as well as the adaptations that they perceived.

Accordingly, the study highlights that much of local adaptive capacity is actually determined by higher organizational (political or economic) levels. The concept of governance recognizes not only that formal decision-making, such as the state and state policy and legislation, may be important in structuring decision-making and the abilities to act within a sector, but that the market system or interest groups may also constitute important influences that constrain or enable adaptations at the local or regional level.

2.1.2. Relation of the study to participatory and modelling-focused vulnerability assessment

In light of the focus described thus far, the study also exhibits important differences to research concerned primarily with direct participation in vulnerability assessment. Vulnerability assessment and integrated assessment have generally focused on interaction with stakeholders that takes place within a participatory framework (Schröter et al, 2005), targeting direct stakeholder interaction, often in groups, rather than, for instance, using interviews or other, more direct data-collection techniques. This framework has often been subject to the requirements of computational modelling, with the end product of the assessment often being presented in numerical terms and as a computer model, with stakeholder information being used to inform the modelling (ibid.). Such model-oriented work often aims at reducing stakeholders' understandings to factors of a higher generality in order, for instance, to develop maps of climate change vulnerability (Patt et al, 2005). In this context, participation of stakeholders is often established in order to increase the policy relevance of the study: for instance, Schröter et al note that a common objective for vulnerability assessment could be 'informing the decision-making of specific stakeholders about options for adapting to the effects of global change' (Schröter et al, 2005, p573).

The focus in the present research is less on informing stakeholders about potential adaptations than it is on understanding in depth the system in which stakeholders act and the constraints on and extent of their adaptive capacity, especially given economic and political changes. In much scientist-organized participation the aim, main concepts and categorizations used are often scientific, and the assumption is that science should teach stakeholders facts about climate and expand their adaptive horizons (cf. Van Asselt and Rijkens-Klomp, 2002). In contrast, a primacy is placed here on stakeholders' knowledge and understanding of the social context that has created the existing situation (and problems) and in relation to which any future information efforts or policy efforts would need to be framed.[1] In this light, a study of social vulnerabilities and adaptation can be seen as a prerequisite for any well-developed efforts at interaction that are to attain such an understanding and organize the enquiry according to stakeholders' rather than scientists' conceptions. Thus, while some participatory elements are used in this study (such as stakeholder meetings/focus groups), the research generally focuses on established social science methodology in terms of individual interviews. This methodology is grounded in an assumption regarding actors' individual and specific knowledge where they are seen as constantly coping and adapting through, for instance, traditional, local or sectoral practices, at least to events within a predictable and known range that is within their capacity, in order to stay in business: what they are doing is 'acting effectively, within their constraints, in their self-interest to reduce their vulnerability' (Chinvanno et al, 2007, p244). Moreover, actors can sometimes be seen to exert political pressure to put into motion changes to raise their adaptive capacity and attain scale benefits – a capacity that differs between groups and actors in different socioeconomic and political (including scale) positions. These power differentials may cause some actors to hold grudges against others or feel reluctant to speak openly about conflicts where people who may adversely affect their resource use are present. As they circumvent such tensions, individual interviews may provide a particularly suitable methodology when studying resource conflicts.

Thus, a qualitative, perception-focused approach could augment the numerical approaches that are commonly used in studying climate change, by supplying a methodology and extensive body of data to guide policy development, the design of information efforts to raise environmental sustainability or agent-based modelling. Such information could also be used to gain an understanding of conflicts and power differentials that may manifest themselves as differential vulnerabilities.

The focus on stakeholder knowledge entails certain qualifications, however. One principal limitation of the material is that, based as it is in stakeholders' understandings, it is limited by any misinformation stakeholders may have been subject to. However, it should be noted that as stakeholders tend to act in accordance with the best of their knowledge, their actions may be the same and have the same effects as if the misinformation were true. In a temporal perspective, stakeholders would also be expected to highlight changes they themselves are aware of. A focus on stakeholders' knowledge would therefore be expected to express changes in relation to the present or near past and future without necessarily being able to relate these to the historical scale or types of change. The study is also limited in taking a sectoral focus, as the stakeholders in a particular sector are likely to highlight the concerns in that context to the exclusion of others; in other words, they do not make the trade-offs with other sectors and interests that public policy, for instance, is to some extent required to make. Thus a sectoral study always excludes some perspectives, for instance those of other sectors whose use of resources conflicts with or otherwise impacts the sectors analysed. To gain an accurate description of vulnerability in a region, several different sectors and a large number of interests would need to be compared and contrasted. This study could have benefited from expanding the study to additional case study areas as well as looking at environmental protection in the areas and nature tourism and, to a lesser degree, at other sectors cited by reindeer herders as impacting that sector. A study such as this could also be extended beyond its present form, for instance through a survey that would apply the specific results here to the sectors at large in the focal areas or as part of the development of a comparative database of cases.

2.2. A structure for vulnerability assessment

2.2.1. Overview of the methodology

In order to fulfil the aims of a vulnerability assessment as described above and in the previous chapter, this study first assesses the present socioeconomic, political and environmental situation of the areas under investigation. This assessment draws on a qualitative multi-method design centred on interviews with individuals and with groups of stakeholders selected by economic sector. As a supplementary source, to triangulate (and improve the reliability of) the material as well as to develop an understanding of the extent to which stakeholders' views on resource conflicts may resonate more broadly with views in the areas, stakeholders' narratives were compared with corresponding accounts in local newspapers, in particular letters to the editor. The study extracted projected impacts of climate change from the natural science impacts and scenario literature and formulated these into concrete statements on impacts, for example on changes in the length, temperature and humidity of the seasons. These statements were presented to stakeholders in interviews and group discussions in order to ascertain their understanding on a concrete level of what specific projected impacts would mean for them. Stakeholders reported both actual ongoing adaptations to situations that resemble those caused by climate change, such as warmer summers, and potential adaptations by saying what they thought they would do if the changes took place (cf. Schröter et al, 2005).

2.2.2. Case study selection

Much of the climate impact literature on vulnerability and adaptability describes how particular 'exposure units' respond to climate change. These units are often geographic regions, countries, sectors, ecosystems or, less frequently, social groups (O'Brien et al, 2004a). O'Brien and Leichenko point out that clusters of countries can be seen as areas that share common characteristics, such as economies in transition or semi-arid environments. The countries in a cluster can also be analysed from a regional perspective. A sectoral perspective is often taken when a particular sector of the economy is considered especially vulnerable. Social group perspectives in turn are applied to especially vulnerable groups within sectors or regions. Natural ecosystems may also be seen as particularly vulnerable to change, with changes leading to the restructuring or demise of local industries, such as fisheries or forestry (O'Brien and Leichenko, 2000). The focus in the present study is on sectors within a region, the rationale being that sectors dependent on renewable resources can best reveal both present socioeconomic changes and environmental changes, including climate change. Community-based vulnerability assessment has also often been applied in practice to rather homogeneous village-sized localities reliant on a limited number of occupations (often subsistence) (Berkes and

Jolly, 2001; Ford and Smit, 2004). For larger communities with greater diversification in types of occupations and livelihoods, as in the present study, such an assessment needs to be modified, for instance by a focus on economic sectors.

This study focuses on case studies in the northernmost regions of Norway (Finnmark County), Sweden (Norrbotten County) and Finland (the Province of Lapland) – a region that can be seen as vulnerable both socioeconomically and in terms of climate change. In the case of climate change, the northern regions of the globe are among those where the impacts will be felt the earliest: strong impacts may be expected where climate change affects the freeze–thaw relation, in other words causes temperature shifts affecting whether systems are frozen or thawing (ACIA, 2005). The region is also socioeconomically and politically vulnerable due to its peripheral location relative to markets and decision-making centres, its sparse and aging population, and the ongoing structural transformation from an industrial to a service economy, which has included rationalizations in traditional resource use sectors such as forestry and fishing.

The present study examines the sectors of reindeer herding, forestry and fishing for a number of reasons. First, they are among the most rapidly changing sectors, as the industrialized, largely natural resource-based economy is becoming increasingly globalized, resulting in an increase in local socioeconomic vulnerability due to the large-scale internationalization of companies and markets for production globally. As sectors characterized by a dependence on renewable natural resources, they are also closely related to the environment and are likely to be impacted not only by economic and political changes but by climatic changes as well. The major sectors of renewable primary production still remain important for the three countries studied here, although a majority of the population in the region now work in the service economy (Pettersson, 2002). In Sweden and Finland, forestry constitutes a major part of national industry exports, accounting for 20 to 30 per cent of the total GDP. Fishing is one of the major export industries in Norway and together with aquaculture accounts for about 10 per cent of net exports (Myrstad, 2000; Boreal Forest, 2005; Gateway to Sweden, 2005).

The third sector examined is the much smaller, but very environmentally dependent livelihood of reindeer herding, which further highlights issues of traditional resource use among various groups. Reindeer herding is practised to a large extent by indigenous people and in the same areas as forestry or other extensive land uses, which has led to major land-use conflicts. The relationship between forestry and reindeer herding is governed by a large body of national legislation and regulation. Where both livelihoods are carried on side by side, the research must address the issue of multiple-use areas and conflicts over the management of natural resources that hamper adaptation to change. The case of reindeer herding is also interesting in that it illustrates how even an occupation that is often thought to be traditional and subsistence-based has been integrated into the world economy: it reveals an increasing international and market orientation in keeping with recent developments in the meat market, meat sales being the main income for reindeer herders. Indeed, the research on reindeer herding conducted here strongly suggests that there are few places today where adaptive capacity can be conceived of as being created and sustained by a single community or the decisions made there.

Forestry will here be studied in Sweden and Finland in areas where the industry is particularly prominent. Fishing will be studied in northern Norway in an area where the livelihood is traditionally practised by local people, who include the indigenous Saami population, which adds an additional level of resource considerations to the issue. Reindeer herding will be studied in all three countries – Norway, Sweden and Finland, which are the three main countries in which indigenous Saami practise the livelihood. In Norway and Sweden, reindeer herding may, with some exceptions, only be practised by Saami. In Finland, in contrast, both Saami and Finnish residents are generally allowed to practise the livelihood, although the two groups tend to be concentrated in different parts of the reindeer herding area. Given these complexities, the study will not attempt to categorize the resource users according to their use of different types of ecological, traditional, indigenous, and/or local knowledges,[2] but will simply note that different actors may possess different types of knowledge and apply this to resource use situations and adaptation. Indigenous groups' rights will instead be discussed within the framework of indigenous norms, in accordance with the theoretical focus on governance. In sum, the study illustrates both local and indigenous resource use within the context of a modern market economy.

2.2.2.1. Case study areas

The selection of specific case study areas in the targeted region was conducted on two bases: the status of forestry and fishing as major occupations and the traditional importance of reindeer herding in the regions. Accordingly, forestry is examined in important forest industry areas in northern Sweden and Finland; fishing is studied in northernmost Norway, where small-scale fishing, including by Saami, has long been prevalent; and reindeer herding is analysed in northernmost Norway, where the livelihood is traditionally strong, and in areas that overlap those of forestry in northern Sweden and Finland. In the case of Finland, the study focuses on an area where reindeer herding is practised predominantly by Finnish (non-Saami) herders in order to provide a comparison with the indigenous practice in Norway and Sweden. The case study areas centre on river valleys, which historically have been important organizational units for the sectors as population centres and, in the case of forestry, also used as channels to transport logs downstream. In practice, the study focuses on relatively peripheral municipalities in these areas, potentially those most vulnerable to change. However, it also includes actors in the larger municipalities in order to represent the major businesses active in the areas and to include administrative actors at the regional level, who play a large part in administering sectoral resource use. All the case studies include municipalities ranging in population from a few thousand (often in inland areas including some very small municipalities of a few hundred inhabitants) to some tens of thousands (often on the coast).

The Swedish case study area in Norrbotten County lies mostly in the valley of the River Pite, which is here defined to include the municipalities of Arjeplog and Arvidsjaur, inland communities with fairly large areas set aside for environmental protection, as well as Piteå and Älvsbyn, which are closer to the coast. All four municipalities are situated in the southern part of Norrbotten, the northernmost county of Sweden. Arjeplog has a population of around 3000, Arvidsjaur around 7000 and Älvsbyn around 8000, about half of whom live in the central towns. Piteå, the largest municipality, has some 40,000 inhabitants (22,000 in the city proper) and has large-scale sawmills and pulp and paper mills belonging to the multinational forest company SCA and, through subsidiaries, also to the state-owned forest company Sveaskog. Norrbotten is also the county in Sweden where reindeer herding is most common.

The Finnish case study centres on the valley of the River Kemijoki, which is situated in the southern part of the country's northernmost province, Lapland. Selected communities in the area range in size from several thousand to 35,000 (in the coastal town of Kemi). The case study area includes the municipalities of Salla, on the Russian border, as well as Savukoski, Pelkosenniemi, Sodankylä and Kemijärvi. Large forestry units are found in Kemi and to some extent in the inland town of Kemijärvi. Reindeer herding is important in the area, which includes the Kemi-Sompio Reindeer Owners' Association, Finland's largest.

The Norwegian case study area centres on the county of Finnmark, the country's northernmost county, with a focus on the valley of the River Tana. The immediate river valley area has been defined to include the municipality of Tana as well as very small-scale municipalities such as Tana Bru, which has only a few hundred inhabitants. Norway is the country where the most Saami live, and Finnmark constitutes a crucial Saami area with a strong reindeer herding component and larger herds than in the other areas; accordingly, important Saami centres such as Kautokeino and Karasjok have been included in the study. Fishing plays a crucial role in Finnmark, which accounts for about 10 per cent of the national fish production (Finnmark County Administration, 2000). Where fishing is concerned, this study focuses on coastal fishing but includes some examples from river fishing as well. To gain a perspective on coastal fishing and administration, the research includes interviewees from larger, mainly coastal towns, for example Vadsø – the provincial capital of Finnmark – Vardø and Alta. Figure 2.1 shows the case study areas and marks the main locations for the case studies.

Figure 2.1 *Map of the northernmost parts of Norway, Sweden and Finland, with main case study locations marked*

Source: Figure courtesy of Nicole Ostländer

2.2.3. Methodology and material

Stakeholders were here seen as those groups with a stake in resource use within a particular sector and area (Mostert, 2002; Jürgens, 2002; Keskitalo, 2004a; cf. Nordic Council of Ministers, 2002). Perceptions and priorities among these stakeholders were mainly studied through semi-structured qualitative interviews using open-ended questions. These constitute most of the empirical material.[3] A total of some 55 interviews (somewhat fewer than 20 per case study area/country) were carried out in the areas with stakeholders in the different branches of the focal sectors between late 2003 and early 2005. Interviewee selection for the study centred on the targeted economic sectors and formal decision-making in the area, such as local government and branches of administration that could be expected to influence adaptation and decision-making (cf. Gissendanner, 2003). The aim was to include a cross-section of the sector and thus maximize variation among the interviewees for the sector and area in order to gain information on the variation of features identified by stakeholders as important for their adaptive capacity. The interviewees thus include both practitioners and administrators, with a focus on the local level. In the case of some regulatory functions that did not exist on the local level, regional-level interviewees were selected. The stakeholder groups included:

- large- and small-scale forestry and fishing companies (active in raw material extraction such as logging or fishing);
- owners/managers of private raw material processing installations in the area (both large and small scale, such as pulp and paper factories and sawmills);
- individual forestry machine operators, fishers and reindeer herders;
- members of private, public and sectoral interest organizations;
- officials in regional administration; and
- officials in local government.

The interviewees were selected through a stakeholder analysis of who the relevant (here sectoral) actors were in the selected areas, including responsible organizations at the regional and local levels (cf. Mostert, 2002), with the aim of representing the main occupational groups active within the sector in the case study areas. Within these groups, a focus was placed on selecting the companies and persons who had been active the longest in the area and would thus be best able to describe change over time. For the focus groups or stakeholder meetings, the number of invited actors in the different categories was expanded in order to assess the extent to which other stakeholders agreed with previous interviewees' statements and the researcher's understanding of these. Maintaining the above criteria, some of the selection of interviewees also took place by snowball sampling, a social science methodology for targeting social networks in which interviewees are asked to suggest other relevant interviewees.[4] A full listing of the interviewees can be found in the references.[5] For reasons of confidentiality and to make it possible to discuss even locally sensitive issues such as resource conflicts or criticism of resource management and organization, interviewees were anonymized and are only identified in the research by profession or organization and general case study area. Large companies or organizational units are an exception here, as the interviewee cannot be recognized even with the reference to the company or unit name.[6]

The individual interviews followed a semi-structured protocol, which allowed interviewees to describe the situation in their own terms while ensuring that they cover the topics needed to assess the present problems, opportunities and possible impacts of climate change (cf. Jones et al, 1999). In the interviews, stakeholders were invited to draw out the implications for their occupations or organizations of specific projected impacts in terms of vulnerability and capacity to adapt (cf. UK Climate Impacts Programme, 2001). The interviews were structured in terms of the following four themes in order to provide a picture of the interviewees' vulnerability:

1. the general socioeconomic situation of the interviewees, including present problems and possibilities, and perceived trends or changes during their working life, as well as the decision-making or governance network which they regarded as affecting them;
2. the interviewees' possibilities to adapt to the changes they identify;

3 the interviewees' perceived sensitivity to specified climatic changes and the environmental changes they identify; and
4 the interviewees' possibilities to adapt to such changes.

Theme 1 served to describe sensitivity to change overall, which interviewees primarily defined in economic and political terms, theme 3 climatic sensitivity, and themes 2 and 4 adaptive capacity in general and capacity to adapt to climate change in particular (cf. Smit and Pilifosova, 2001; Smit and Wandel, 2006). Thus, under theme 1, the interviewees were asked to describe their everyday work, what factors have an impact on this work, whether these factors have changed since they started work, and what the present problems and possibilities in their occupation and everyday situation are. They were in general not guided specifically towards economic or political themes, but were allowed to speak freely; the economic focus articulated by many of the interviewees thus indicated the importance of economic considerations. Under theme 2, the interviewees were asked in what way their situation could be improved and what means they had available to adapt to changes in it. Under theme 3, specific projected impacts of climate change that had been derived through a survey of climate and impacts literature were mentioned using questions such as 'How would it impact you if the winters started later?'. Interviewees responded by mentioning the outcomes and sometimes also existing changes, thereby indicating actual rather than merely potential courses of impact and adaptation. Finally, the interviewees were asked how they could adapt if these changes occurred, in response to which interviewees at times discussed present adaptation to such events.

Thus, a lay-language phrasing of climate change impacts, rather than the term 'climate', was used in the communication with stakeholders. This approach was chosen because the issue of climate change may carry specific assumptions and non-scientists may respond to the terms on the basis of what they have read in the media or misunderstand scientific concepts unless they are defined and presented in lay language (cf. Bostrom et al, 1994). For instance, Petts et al suggest that participation must employ the language of the user, which means not only using 'plain language' but attempting to understand how people interpret key phrases and concepts and what they know about science in order to adjust communication accordingly (Petts et al, 2001).[7] This suggests that interaction with stakeholders must not assume that 'climate change', even if well publicized, is a familiar and well-understood concept or that people have a well-developed understanding of how climate change would affect them. In order to avoid interviewees talking about how they understood the concepts rather than about actual concerns and their anticipated impacts, as a rule the study aimed to use neither the term 'climate change' nor 'globalization' in interviews or interaction with stakeholders. Additionally, given the focus on adverse impacts in some social vulnerability research (cf. Kelly and Adger, 2000), the present study formulated impact-related questions without an assumption of negative changes by asking questions such as

'What changes have you seen in your working life?' and 'How can you adapt to these?'. In practice, however, much of the discussion concerned possible adverse impacts (reflecting stakeholders' concerns). The interviews were transcribed in full, and the excerpts quoted in the text are the author's translations.

The research process further included feedback and verification of the reliability of the interview results through focus group interviews with actors (including both previous interviewees and additional actors in the targeted stakeholder groups). These group discussions were held to provide forums for direct dialogue and were organized to take up open questions and discuss vulnerabilities, adaptive possibilities and cooperation. They were also used to check the results of observation and interviews and to provide feedback to the communities on the progress of the study (cf. for example, UK Climate Impacts Programme, 2001; Gregory et al, 2003). The format of the discussions largely followed the recommendations for integrated assessments of climate change with focus groups: the groups included six to eight persons in a workshop format and met at neutral places, and the output of their work consisted of audio tapes and minutes. The role of the researcher was that of moderator or discussion facilitator (cf. Dürrenberger et al, 1999). In total, seven such focus group meetings were conducted – two or three in each case study area, not including the shorter feedback sessions with individual stakeholders – and they largely supported and in some cases refined the understanding of the interview data.

While the emphasis in the work is ultimately on the interview material and, to a lesser extent, the focus groups, the study includes as a supplementary source a newspaper survey based on a selection of the regional papers in the areas.[8] The survey consisted of electronic searches in the databases of those publications for the previous five years (2000–2004), as well as manual searches over shorter periods of time in cases where material was unavailable in electronic form. The searches centred on key words in the areas of forestry, fishing and reindeer herding, and yielded a body of material – including many letters to the editor – that indicated substantial public involvement with certain issues targeted by stakeholders and lent a more general currency to the stakeholders' views. The newspaper survey was used for a background triangulation of the results of the interviews, in other words to check their applicability to broad local/regional discourse and especially to provide a context for contentious issues such as conflicts over land-use rights discussed in interviews and meetings. The survey was used to indicate whether these issues were discussed more broadly in the study areas in the manner suggested by the interviewees' statements. The survey is occasionally referred to in connection with some of the more contentious or conflict-oriented issues discussed, but is otherwise not dealt with in the work.

2.2.4. Identification of governance networks and norms

Coding refers to the process of organizing data into categories. To ensure exact documentation of interviewees' statements, interviews and stakeholder meetings were tape recorded and transcribed verbatim, with the subjects' permission (cf. Kempton, 1991; Marsden, 1997). The data was then coded using, among other tools, the Atlasti coding software for qualitative material analysis. Coding was undertaken to include all statements that referred to factors of change (Dürrenberger et al, 1999; Gregory et al, 2003) and adaptation to these changes, as well as the governance context for the adaptation. This approach accorded with the stated aim of covering both general and specific climate change vulnerabilities and achieving a maximal variation among interviewees with regard to the features that define their adaptive capacity in response to perceived changes. These features are systematized under larger themes (such as 'Rationalization and technological change'), which are dealt with under the major themes of economic, political and climate change. The material is presented in the form of extensive qualitative narratives with a substantial focus on illustrative quotations and examples.

Networks of governance were defined here in terms of the perceptions by local and regional actors of *which actors and mechanisms on what levels* influence their abilities to act. This echoes similar observations by Cannot et al that power relations have to be 'traced back from the immediate assets and livelihood base of a household along a "chain of causation" back to the processes and institutions that determine the distribution of safety and vulnerability in society' (Cannon et al, 2003, p6).[9] Accordingly 'governance' refers to the decision-making structure as seen by stakeholders and the degree to which this structure includes actors at the international, European Union, national, regional and local levels, administrative, policymaking and private actors, or interest groups. Descriptions of the governance framework will serve to highlight the level at which decisions relevant to the local level are taken. A governance network approach may thus indicate the levels at which some of the adaptive capacities and suggested adaptations to change lie, one implication of this being that vulnerability may be decreased or increased by actions at different levels. Drawing upon the approach by Ziervogel and Downing (2004), governance networks are represented graphically in terms of the different scales of decision-making.[10] Figures in Chapters 3–5 illustrate the location of different groups of actors – administration and policy, industry, and interest groups (cf. Bakker et al, 1999) – at the local, regional, national and international levels. The figures depict the governance network for each sector as the interviewees perceive it, indicating the levels and types of actors seen by interviewees as impacting their local situations.

The concept of norms is used to distinguish the international values that stakeholders highlight. In practice, these often represent norms that stakeholders aim to enforce or dispute. These norms and the processes that support them have been mentioned by interviewees in response to questions concerning their everyday resource use. The types of norms discussed by the actors mainly concern environmental and indigenous peoples' rights:

- *Environmental protection norms* are visible in international and European Union (EU) regulation and state environmental protection legislation and policy. The development of the framework for environmental protection is perceived by stakeholders as having had a strong influence on conditions within the sectors.
- *Norms on both environmental protection and indigenous peoples' rights* are visible in forest certification as a market-driven, formally voluntary labelling system for environmentally and socially responsibly produced wood that includes environmental requirements and requirements in relation to indigenous peoples.
- *Norms on indigenous peoples' rights* are manifested in Convention No 169 of the International Labour Organization (ILO) on indigenous peoples' land rights. Norway has ratified the convention, while, at the time of the study, Sweden and Finland were investigating potential ratification.

Actors in forestry in particular mentioned the impact of environmental norms that have been institutionalized in environmental protection and forest certification measures and, to some extent, norms on indigenous peoples' rights as constraining their local resource use. Indigenous actors in reindeer herding and fishing instead highlighted the empowering element of norms on indigenous peoples' rights. There may, and probably do, exist parallel norms that can be seen as creating and exacerbating conflicts that then play out domestically or through international networks and organizations; the focus here will be confined to the cases discussed by interviewees as affecting their behaviour and scope of adaptation.

2.2.5. Synthesis of climate change scenario and impacts literature

To make it possible to access data on people's vulnerability, adaptive capacity and specific adaptations in response to climate change, interviewees were presented lay-language statements relevant to their occupations. The survey of the literature on climate change and impacts with regard to northern Europe and the northern parts of Norway, Sweden and Finland in particular includes international (IPCC), regional (IASC), national (SWECLIM for Sweden, RegClim for Norway and SILMU for Finland) projections, as well as the projections found in international and national research programmes and the impact and scenario literature in general. Drawing on these studies, the author has derived the changes that are particularly relevant to the forestry, reindeer herding and fishing sectors. Interviewees were asked to describe how these main projected changes would impact them and how they could adapt. Below are the main projected changes for each of the sectors and the possible impacts that were presented to interviewees. To exemplify the climatic condition of the areas in question, Finnish Lapland has a winter season ranging from first snow at the end of October (before which the ground has usually frozen) and snowmelt in late April or early May. The average length of the growth period (daily mean temperature >+5°C) is between 100 and 140 days (Kumpula, 2001; cf. Rikkinen, 1992).

2.2.5.1. General impacts of climate change

Any assessment of future climate change includes uncertainty: assessments are built on assumptions on future population, energy use and technology; these may be changed by as yet unknown parameters of how systems interact and feedback into each other, and more general trends are likely to manifest themselves differently at different locations depending on local conditions. Any assessment of climate change is therefore uncertain and includes considerable possible variations (Smith et al, 2001). In general, climatic changes are already ongoing, and can be seen in the fact that the growing season in Norway, Sweden and Finland lengthened between 1890 and 1995 and that recent warming trends have already been noted in the North Sea (Smith et al, 2001). Scenarios developed by the IPCC indicate temperature rises up to 2050 of about 3°C in mean winter temperature and about 2°C in summer compared to 1990 values (Sweden, 1997). The Swedish Regional Climate Modelling Programme (SWECLIM), the Regional Climate Scenarios for Norway (RegClim) and the Finnish climate impact study SILMU each predict that the warming trend will be strongest in winter and in inland and northern areas, although their specific predictions differ (Høgda et al, 2001; FINSKEN, 2003). Precipitation is expected to increase by 10–15mm, with the greatest increase in the winter months (Sweden, 1994; Sweden, 1997); a Finnish study indicates that precipitation could increase by as much as 25 per cent by 2050 (FINSKEN, 2003).

For the three countries as a whole, a largely positive net impact of climate change can be expected at the national level, although the size of that impact is still unclear (Kuoppamäki, 1996a; cf. Forsius et al, 1996). According to some calculations, Finland's economic benefit from climate change will be slightly over FIM4000 million (US$800 million), which is as much as the country earns from forestry, or twice what it receives from agriculture (benefits in 2050 expressed in 1993 money). The figure is about 1 per cent of the 1993 GDP. The projected benefit derives mainly from increased productivity in agriculture and forestry, and savings in energy. The highest costs come from, among other things, the anticipated reduction in biodiversity (Kuoppamäki, 1996b). According to the Swedish report to the IPCC, the sensitivity analysis for technical systems and the possible effects of a milder climate show that no significant changes will occur in the short term that might entail a substantial societal risk (Sweden, 1997). The impacts of climate change on employment in the fisheries of northern Norway are projected at around ±1 per cent of the total employment in the area, depending on whether biological productivity increases or decreases. However, varied management practices yield different impacts at different locations depending on the socioeconomic structure of the affected community and how fishing is organized there (Lange, 2001).

Measures have been taken at the national level in response to anticipated climate change to limit emissions, promote and coordinate research, and inform people about the issues, but it is not possible to say how these actions will affect the numbers and estimated impacts (cf. Sweden 1994 and 1997; Norway 1994a and 1994b; Finland 1995a and 1995b; Huq et al, 1999; Ouedraogo et al, 1999). It is also not possible to form a general viewpoint that is more than a rough estimate of how the trends for the countries as a whole will impact specific regions.

The maps below illustrate the projected temperature change as of July 2020 and 2050 in degrees Celsius as compared with the temperatures in 1990. The climate change models on which the data are based indicate that a territorially differential and accelerated increase in temperature can be expected. Indeed, the projected changes may result in very different effects on different scales and in different localities or regions, making it possible that local vulnerability will significantly differ from national assessments.

In the case of projections, this study refers to broad trends that could be of different magnitudes but that say something about the parameters that could change; this makes it possible to investigate the effects and adaptive capacities which the interviewees have seen. In general, the projected changes in temperature and precipitation will result in a more temperate climate with delayed autumns and milder winters with increased precipitation (Finland, 1995b; Høgda et al, 2001). According to RegClim (Sygna and O'Brien, 2001), the winter season will be shortened by perhaps a month. River and lake ice will form later and break up earlier (IPCC, 1998). What is especially important in these areas is also the freeze–thaw threshold of 0°C. Any climate change that shifts the freeze–thaw line, whether in space or time, will bring about important impacts on natural systems (IPCC, 1998). This means that climate change that causes winter thaws or re-freezing following a thaw in spring may have significant consequences. The likelihood of such events may increase, given that climate change is expected to lead to greater variability in the climate in the form of more frequent storms or other extreme weather events (Sweden, 1997; Sygna and O'Brien, 2001). In general, however, autumn will start later and winters will be warmer with increased precipitation.

The impact of climate warming on the onset of spring is dependent on several factors. Two scenarios have been posited. In one, increased precipitation and temperature will result in an earlier onset of spring in areas where snow has been blown away by stronger winds; spring will come later in areas with a thicker snow cover, where the snowmelt will take longer (Høgda et al, 2001). In the other, the start of the snow cover season will occur later in the autumn, and melting will start earlier, shortening the winter season overall. Saelthun suggests that the total snow

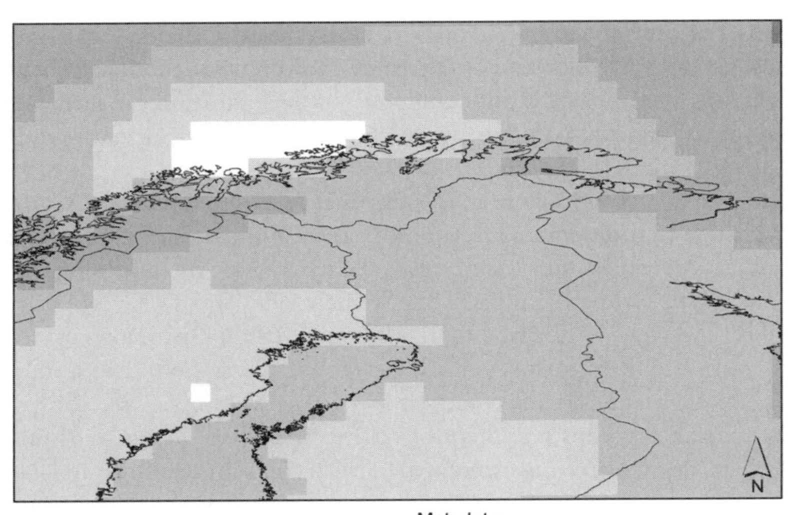

Temperature Change in July 2020 in °C

Metadata:
Model name: REMO
Model author: Max-Planck-Institute Hamburg
Data processing: Mean temperature change for the years 2016-2025

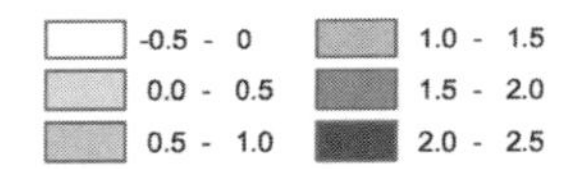

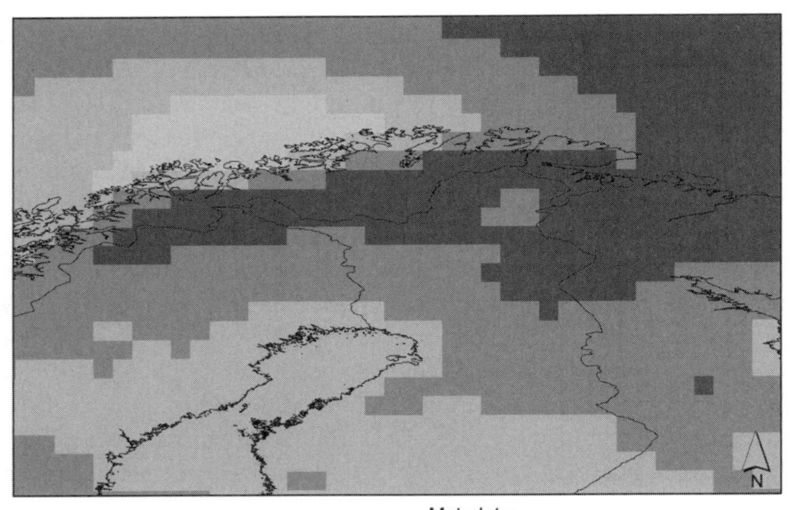

Temperature Change in July 2050 in °C

Metadata:
Model name: REMO
Model author: Max-Planck-Institute Hamburg
Data processing: Mean temperature change for the years 2046-2055

Figure 2.2 *Maps illustrating projected average temperature increases for 2020 and 2050*

Source: Courtesy of the EU-funded BALANCE project, project number EVK2-2002-00169

depth and the snowmelt run-off will be reduced and that the reduced snow cover will result in earlier snow-free ground (Saelthun, 1995; cf. Guisan et al, 1995). The spring flood will be affected accordingly and the time and peak of the flood may change (IPCC, 1998; Høgda et al, 2001).

In summer, plants may become stressed by heat, a deficit of soil moisture due to the longer growing season and changing precipitation patterns, and in-migrating pests, pathogens and herbivores (Norway 1994b; Saelthun, 1995). On the other hand, growth conditions may improve as a result of warmer temperatures and the ground being free of snow for a longer period. Some studies have indicated that increases in temperature over the last century have in fact begun to create such impacts. A change in climate – or at least changes posited as resulting from such a change – can thus already be seen in some instances (Sygna et al, 2004), meaning that impacts and adaptation to climate change may already be ongoing (as some interviewees in fact indicated by describing existing adaptations to what were framed as projected trends).

Interviewees were asked, in general, to react to statements about the impacts described above. Local stakeholders were asked how it would impact them if the onset of spring varied or came earlier, summers were warmer, autumn came later, and the winters were milder, with possible thaws. They were also asked about the changes they had perceived over time; often, interviewees responded that they had already seen some of the changes, such as warmer winters. Impacts specific to each sector (forestry, fishing and reindeer herding) were also investigated with reference to the primary resource.

2.2.5.2. Impacts of climate change on forestry

One of the most important impacts for forestry will be a longer growing season and consequent increased forest growth (Lee, 1999; Høgda et al, 2001). This will result in increased volumes of timber but reduced quality vis-à-vis the slow-growing, strong wood with thin and dense growth rings that is currently typical of the focal areas (Layton and Pashkevitch, 2000). Increased temperatures may also result in better seed crops and germinability (reproduction) (Molau, 1995). In principle, warming, a longer growing season, an increase in CO_2 levels and added humidity should have a positive effect on the growth and reproduction of forest trees; at present, growth cycles in these areas are long, with trees usually harvested at 90–110 years of age. These positive effects may be offset, however, by leaching of nutrients from the soil due to excessive rainfall, disturbances in winter dormancy, frost damage due to sudden cold spells in spring, storm and snow damage, new plant diseases and pests, and more frequent forest fires (Finland 1995b; Kuoppamäki, 1996a, 1996b). In addition, heavy snowfalls and ice formation may increasingly crush tree crowns and new growth (Layton and Pashkevitch, 2000).

Changes in climatic conditions will impact the distribution of and competition among tree species. Over several hundreds of years, the present-day dominance of

Norway spruce (*Picea abies*) and Scots pine (*Pinus sylvestris*) in the areas may shift, yielding to mixed forest with more Scots pine, as Norway spruce is expected to grow poorly in a milder and more maritime winter climate (Sweden, 1994; cf. Guisan et al, 1995; Skre, 1999). Terrestrial ecosystems are also likely to change from tundra to boreal forest (IPCC, 1998), and the climatic timberline may be elevated due to increasing temperatures. Over the shorter term, however, the increase in forested area will not be very large; any benefits to the sector will accrue from increased growth rates (Holten, 1995; Layton and Pashkevitch, 2000). Forestry will also be impacted by changes in accessibility to roads and logging areas: for example the seasonal thawing will reduce accessibility and impede transport (Sweden, 1997). Climate change could also result in disruption and increased maintenance costs from ground subsidence, landslides and icing (IPCC, 1998).

To accommodate these projected changes, the interviewees in the forestry sector were asked if, and, if so, how, increased forest growth, a longer growing season, possible thawing and re-freezing in spring, possible increases in snowfall, and a shift in tree species and spread would impact their livelihood, including their use of roads.

2.2.5.3. Impacts of climate change on reindeer herding

Reindeer will be significantly affected by climate change through impacts on grazing. Reindeer graze mainly on lichen in winter, with their diet including arboreal lichen species as emergency fodder (where these exist); in summer they eat grasses and plants. In a warmer climate, lichens and mosses will be replaced over the longer term by transpiring plants and shrubs (Guisan et al, 1995; Henttonen, 1995; IPCC, 1998). More immediate impacts may be felt in the several periods during the year when environmental and weather conditions pose problems for reindeer and reindeer management. In winter, thawing or rain that result in ice crusts on the snow may make it difficult for the reindeer to move around and especially to dig through the snow and layers of ice to access lichen (Sweden, 1994; Henttonen, 1995). Grasses important for early spring grazing – when fodder is running out after a long winter – may be impacted by spring frost (Guisan et al, 1995; Henttonen, 1995; IPCC, 1998). Summer grazing in turn may be restricted by adverse grass and plant growth conditions, a lack of water and the presence of biting insects, which disturb reindeer (Heal et al, 1998). Negative conditions in these respects increase the animals' energy expenditure and reduce their food intake, resulting in reduced growth and fertility. At present, however, reindeer are fed during certain periods in essentially all reindeer management areas, which reduces the interaction between the animals and pasture resources (Kumpula, 2001).

To account for these factors as comprehensively as possible, interviewees in reindeer herding were asked if, and, if so, how, increased thawing and re-freezing in winter and spring, warmer summers, and changes in lichen and other grazing accessibility and spread over time would impact them.

2.2.5.4. Impacts of climate change on fishing

Water temperatures directly affect the productivity of fish species and the relative abundance of different species. Warming in high latitudes should lead to increased total biological production, longer growing periods, increased growth rates and, ultimately, increases in the general productivity and volume of fish (IPCC, 1998). This would enlarge catches, as species such as cod, herring and capelin could increase their populations to a level comparable to historic maximum values, or some three times what they are today (Klungsøyr et al, 1995). The fisheries may expand, with a general northward shift in the distribution of fish stocks (Norway, 1994b; Klungsøyr et al, 1995). Large areas in the north and east may be colonized by cod, haddock and herring. More southern species, such as mackerel and sardine, may also migrate into northern areas. Spawning grounds that are sensitive to temperature changes may, however, be negatively affected, and nutrient levels may decrease due to warmer temperatures, reducing fish populations accordingly (IPCC 1998; Sygna and O'Brien, 2001). Increased summer temperatures could result in an increase in pests or algal blooms, affecting some fish species (Sygna and O'Brien, 2001).

There may also be greater variability in the climate, seen for instance in a higher frequency of storms or other extreme weather events (Sygna and O'Brien, 2001). However, fisheries are also subject to extensive short-term variations that are not related in any obvious way to climate conditions or human activity, and the great uncertainty of both prognoses and possible impacts should be stressed (Knapp et al, 1999). It has been speculated that increased melting of the polar ice and the resulting higher proportion of fresh water at the surface might impact the flow of water into the Atlantic Ocean, causing the temperature in the northeast Atlantic to decline and resulting in collapses in fish populations (Klungsøyr et al, 1995).

Interviewees in fishing administration and the fishing industry, and fishers themselves, were asked primarily how the most frequently discussed potential impacts on fisheries would affect them. The set of statements presented included if and how they would be impacted by warmer water temperatures, extensive variations in weather, and changes and geographical shifts – mainly northwards – in species, with possible increases in fish biomass and growth.

2.2.6. Understanding change

The vulnerability assessment framework for this study focuses on economic, political (including governance and normative) and climatic factors of change, with a particular emphasis on describing multi-level impacts and multiple causes of change that impact the overall vulnerability of areas. The descriptions in the sectoral case study chapters (Chapters 3–5) endeavour to provide 'thick descriptions', in other words the interviewees' own words about their situation and how they experience it (cf. May, 2001). The work centres on describing broad trends and interactions between important elements in light of interviewees' perceptions and priorities.

The three following chapters examine vulnerability in the forestry, reindeer herding and fishing sectors respectively. The chapters describe the overall situation and changes perceived by interviewees under headings of change in economic, political and climatic conditions, as well as the actors on different levels that stakeholders view as making up the governance system that determines their local 'adaptation space' (Berkhout et al, 2006). Each of the case study chapters concludes with a summary of stakeholders' perceived vulnerability and potential (and, in some cases, actual) adaptations in the sector.

3

Perceptions of Change, Vulnerability and Adaptive Capacity among Forest Industry Stakeholders in Northern Sweden and Finland

3.1. Introduction: The organization of forestry in Sweden and Finland

Forestry is one of the most important resource industries in Sweden and Finland. About 50 per cent and 60 per cent, respectively, of the surface area of the two countries is productive forest, most of which is situated in the north (Sweden, 1994; Sweden, 1997; Finnish Forest Research Institute, 2002). The countries are thus 'forest giants' (Boreal Forests of the World, 2003a; cf. Huq et al, 1999). For instance, in Finland, the forest sector accounted for 27 per cent of the total value of exports in 2001 and its share of the gross domestic product was more than 7 per cent (Finnish Forest Research Institute, 2002). The dominant tree species in the region are Norway spruce and Scots pine, with birch occurring to a lesser extent (Sweden, 1994).

While forestry in Sweden and Finland exhibit certain differences, the sectors in the two countries also show some important similarities. One general similarity is in the pattern of ownership, whereby a relatively large proportion of the forested land in the two countries is privately owned and state ownership is more extensive in the north than in the south. In Sweden, the state today owns only a small percentage of the productive forest lands, and about a fifth of the total forest land. Approximately half of the country's forests are family owned. Company ownership also plays a large role (Boreal Forests of the World, 2003b; Swedish National Board of Forestry, 2004). In Finland, the state owns some 30 per cent of the forested area, with approximately 60 per cent privately owned (Rikkinen, 1992; Finnish Forest Research Institute, 2006). In the north, however, the state is often the majority owner, in particular for forest in sparsely populated areas. Accordingly, state forest policy has a considerable impact in the areas (cf. Rikkinen, 1992). The state organizations which manage forests in the countries are Metsähallitus (the Finnish forest and park service) and Sveaskog[1] respectively. In both countries, forest policy at the regional level is administered through county or district forestry boards.[2] In Finland, these are directed by Forest Centres, which include representatives from a range of stakeholders in forestry. The agencies which coordinate land-use administration and planning and implement state policy at the regional level in general are the regional councils or administrations: in Finland, the Regional Council of Lapland (Lapin Liitto), in Sweden, the County Administrative Board (Länsstyrelsen).

Finland and Sweden also have the common characteristic of 'forest commons' or 'common forests', which make up several per cent of the total forest area. These forests, often historically established, consist of the jointly managed properties or parts of properties of private forestry owners in some areas. The proceeds from forestry in these areas may be paid back to owners directly or distributed in the form of subsidies for silviculture or planting in the owners' separate forest holdings. Many of the family or smallholder forest owners are organized into forest owners' associations such as the Norrbotten County organization Norrbottens Läns Skogsägare (subsequently renamed Norra Skogsägarna following a merger), which works under the Swedish Federation of Forest Owners (Skogsägarnas Riksförbund) (Boreal Forests of the World, 2003b). Another important category of forest owners in both countries is forest companies. In Sweden, the multinational forest company SCA operates large, modern pulp and paper production facilities in the Piteå area, and the company is among the global leaders in its field. In Finland, the forest company Stora Enso is a particularly large actor, operating pulp and paper mills in Kemi and a pulp mill in Kemijärvi.[3] Over the past 20 years, both SCA and Stora Enso have internationalized their operations, establishing themselves firmly in other European countries and on other continents (Boreal Forests of the World, 2003b).

This chapter presents stakeholders' perceptions of changes in forestry in a framework of socioeconomic and political changes that are broader than, but can be related to, globalizing impacts. It also describes the impacts that stakeholders have seen and anticipate as a result of observed and projected climate change. A final section summarizes stakeholders' vulnerability and adaptive capacity in forestry in the areas in relation to these changes.

3.2. Socioeconomic change in the industry

Stakeholders described a large structural change in forestry over the last 20 years. This has mainly taken the form of increased rationalization, decreased employment at the same time as the timber yield has been increased, increased competition from abroad, a shift from employed forest workers to hired contractors, and enormous technological development. The view of the forests as a resource has also broadened to encompass other main uses than timber, such as tourism and nature protection. This development has contributed to a more critical view of forestry as employment declines regionally while most of the remaining administration is relocated to larger centres, sometimes outside the region.

In general, interviewees perceived the situation of forestry as one of increasing economic internationalization and of a sector facing increased demands. This was seen in companies moving, or being under threat of having to move, if conditions are not conducive to business and in the local situation being at risk from outside forces such as imports and cheaper labour. Another manifestation of this trend is that forestry is no longer municipality-bound; rather, employment is moving out of the community. This in turn has placed higher demands upon contractors, who have taken the place of permanently employed personnel, to be more efficient and to work in larger units, where they can use, for instance, logging lorries more effectively, decreasing the cost to the employer. Additional repercussions include the increasingly time-critical nature of business, extensive import of timber from Russia to Finland and the heightened emphasis on ships as a means to transport timber to its destination. There was also an awareness among local actors of the different growing conditions for forests in other countries. These factors are discussed in more detail below.

3.2.1. Rationalization and technological change over time

The rate and magnitude of change can be seen in the work of both private corporations and the two state management organizations/companies Metsähallitus and Sveaskog; interviewees in the latter noted the reduction in the number of forest workers and administrators. For instance, in the municipality of Savukoski, the number of people employed by Metsähallitus has decreased from 100–150 persons in 1985 to some 20–30 persons today (Local Government, Finland; Metsähallitus, planning): 'I think there are 99 forest workers in the Eastern Lapland region today, working in five municipalities' (Metsähallitus, planning).[4] In Sweden, in 1977 Sveaskog's predecessor had some 150 workers in an area somewhat larger than the one which it now manages; the company now has about 30 workers (Sveaskog, Sweden). Corresponding cuts have been made in private companies; for instance, SCA, the largest forestry company in northern Sweden and today a global company represented in some 60 countries, had four different forest administrative divisions with 30 administrators in each before one early rationalization in 1978; today there is one administration division with some 30 employees (SCA, Sweden).

Despite the reduction in employment – of some 80 per cent of all workers in some cases – the timber yield in both countries has been maintained or, more often, raised (Metsähallitus, planning; SCA, Sweden). The main structural changes in the sector are connected to technological changes. There has been ongoing mechanization since the 1960s. In 1979–1980, log floating was discontinued in the Swedish case study areas (Sawmill Owner, Sweden; Swedish Wood and Tree

Trade Union). In the early 1990s, auto-manual workers or loggers with chainsaws were replaced with forestry machine operators (mainly contractors) operating harvesters, which resulted in large lay-offs (Forestry Machine Driver, Sweden). In the Finnish case study area, log floating continued until 1991; its discontinuation was a change much discussed by interviewees. As one actor said, 'the biggest change was when log floating ended. It was a major change in addition to mechanization' (Common Forest I, Finland).[5] The transition from floating timber to transporting it by lorry was considered beneficial in that it made forestry a year-round occupation. Previously, the bulk of the harvest took place in the winter and was brought to the riversides for flotation and storage (Metsähallitus, management).

The tremendous technological development surprised some of the actors on the ground: 'When I started in 1977, I could never imagine logging by machine' (Sveaskog, Sweden).[6] While early machines often froze and broke, making it difficult to cut 20m^3 in two shifts, today it is possible to cut 70m^3 in three shifts (18 hours) (Forestry Machine Driver, Sweden; cf. Transport Entrepreneur, Finland). One interviewee remembers when the first hydraulic cranes to load wood for transport were demonstrated: 'Old lorry drivers said you couldn't do anything with a hydraulic crane and the reason was that it left such ugly tracks [on the ground]' (Transport Entrepreneur, Finland).[7] 'Now one man can do an incredible amount of work in a day with a crane. You'd need a hundred men to match it' (Transport Entrepreneur, Finland).[8]

These developments have resulted in a very time-oriented business where large log depots are no longer kept but, rather, wood is transported on a 'hand to mouth' basis (Transport Entrepreneur, Finland).[9] Permanently employed forest workers have been replaced to a large extent by contractors. Forest workers become contractors and employers themselves, taking on more responsibility. Having fewer but larger contractors reduces the number of parties that forest companies need to negotiate with and thereby their administrative requirements (Transport Entrepreneur, Finland; Metsähallitus, planning; Forestry Machine Driver, Sweden). This has a downside for local employment, however, since it is quite possible that contractors come in from the outside to cut local forest (Metsähallitus, management; Sveaskog, Sweden). There is also pressure to use equipment as efficiently as possible, leading to demands on contractors to increase the size of their enterprise, to provide more of the equipment themselves and to cut costs in order to compete with other firms offering the same services (Transport Entrepreneur, Finland).

The period has also seen structural changes in forestry ownership and interest in the sector. Many of today's forest owners are pensioners. New owners tend to live elsewhere, there are more women among them, they own smaller plots, and they are less active in forestry in general, possibly holding other, less production-oriented values than previous forest owners. It is also becoming more difficult to recruit workers for forestry. This perceived decline in young people's interest in working in forestry is a problem, particularly since ready labour is crucial for thinning operations and forest management. Harvesting has become mechanized, but it is not possible to mechanize thinning to the same extent; indeed, the labour intensiveness of this work is perhaps one reason why thinning operations and forest management have sometimes lagged behind harvesting in priority (Lapland Forest Centre, Finland; Forest Owners Union in Northern Finland; Common Forest II, Finland).

3.2.2. Impacts on smaller actors: The example of small sawmills

The increased competition over time has also resulted in a substantial marginalization and final removal from the market of smaller sawmills in particular. Sawmilling is seen as a particularly vulnerable industry as it requires large capital investments and steady access to timber resources (Local Government, Finland). Given an increasingly competitive market, where timber resources, for instance, became more expensive in Norrbotten in the 1990s, and heightened demands for efficiency generally, many small sawmills have closed down. Those that survive are 'the really big and automated sawmills' (Local Government, Finland);[10] the 'smaller inland sawmills, they have no chance to compete' (County Forestry Board, Sweden).[11] During the time that one of the interviewee's sawmills has been in operation in the Swedish case study area, from the 1960s, six sawmills have closed down in its vicinity (Sawmill Owner, Sweden). In the companies that remain, competitive efficiency has been increased by large-scale mechanization and by streamlining and concentrating administration in larger and fewer units; these measures have made for less labour-intensive practices (Metsähallitus, planning; Forestry Machine Driver, Sweden).

The increasing competition manifests itself in a need to increase sales (preferably export sales, in order to reach larger markets), cut costs, and still be able to accumulate enough capital to modernize and invest in technology (Sawmill Owner, Sweden). The financing situation is considered particularly difficult by sawmills, as they are small actors with large investment needs but lack the sort of resource backing that larger companies receive from their parent company (Älvsbyhus, Sweden; Sawmill Owner, Sweden). Some of the actors viewed the situation as the 'banks run[ning] the business': sawmills were often closed down on account of:

> *... a lack of liquid assets ... you fall into some kind of situation where your profitability is limited and you try to expand and increase your volume and then the timber becomes too costly and your profitability falls even more and the bank just wants to take its fingers out of the pie as soon as possible.* (Sawmill Owner, Sweden)[12]

An additional problem for sawmills has been that they generally do not have long-term agreements for raw materials from forests. Having only short-term contracts, they do not know at what volume or price level they will obtain timber in one year's time, which may make them reluctant to invest (Sawmill Owner, Sweden).

Export and a focus on external markets are seen here as a way for companies to adapt and strengthen their competitive advantage. For example, the owner of an inland sawmill that has been in business for some 40 years saw the reason for the mill's survival in the start-up support it obtained in the 1960s (which limited early loans), in the support it received from its parent company in later harsh times and in its having continuously been export-oriented. The export orientation, to a large extent a result of the owner's personal interest in working abroad and travelling, has meant that the sawmill products have reached a larger market and that the business has been able to access buyers willing to pay more for the particular wood qualities, thus increasing added value. This sawmill thus exemplifies a close connection in a single individual between production aims and market demands (Sawmill Owner, Sweden).

3.2.3. The mobility of larger actors and dependence on resource access

In the present situation, where forestry is no longer bound to a particular municipality, one principal concern is that the industry will move out of the region. To avoid this, there has been an emphasis on the need to ensure the supply of raw material and to improve the transportation infrastructure (Lapland Regional Council, Finland). Actors showed considerable awareness of the competition from elsewhere. An interviewee working in a common forest in Finland pointed out that in Uruguay a tree matures for harvest in 7 years, as compared to 90 years in the region studied here. 'In Estonia and Southern Finland an aspen can be grown in about 25 years' (Common Forest II, Finland).[13] Actors feared that 'it's not profitable or worth the effort to grow birch slowly here' (Common Forest II, Finland).[14] On the other hand, they also emphasized that slow-growing wood may have a quality advantage (SCA, Sweden).

In the Swedish case, the availability of local timber was a crucial factor for the interviewees. The limited availability of timber today was seen as a result of current forest management practices, conservation and historical factors. The situation has also resulted in an awareness of increased competition and the vulnerability of forestry in the area. During the 1990s, for example, there were discussions about whether to invest in or shut down the SCA wood-processing facilities in the Piteå area, whose equipment was becoming obsolete. Ultimately, the choice was to invest in new equipment, ensuring SCA's presence in the area at least for the lifetime of the equipment, but the situation increased awareness of the importance of making local timber resources available. For instance, one interviewee noted: 'High-capital industry in Piteå alone has invested some [SEK] eight billion and it is not a given that they will stay there; they can move to where the raw material is' (County Forestry Board, Sweden).[15] The interviewee also observed: 'They import timber from the Baltic [countries] today [and] about a million cubic metres from [the neighbouring Swedish county of] Västerbotten ... that is certainly a sensitive issue, not least for local politicians' (County Forestry Board, Sweden).[16] The resource situation has thus resulted in increased imports: 'The shortfall of pine timber today is 1.1 million cubic metres ... and if that is brought in from Russia or Finland or Västerbotten or wherever it makes no difference' (Älvsbyhus, Sweden).[17] Imports are, however, limited on account of the price: 'We have imported very little from Finland, because the competition [with] Finnish sawmill owners is too severe' (Älvsbyhus, Sweden).[18]

This mobility was also emphasized among larger companies. Private companies highlighted their mobility to a large extent, including discussion of possible impacts on the region. The SCA interviewee stated that they are rather 'global in the company',[19] with operations in some 50–60 countries, 'so of course the Swedish market is very small to us' (SCA, Sweden).[20] This international nature makes availability of the primary resource an especially crucial question: 'If it gets too tough and too much of a struggle then there are other places, so why should you beat your head against the wall here' (SCA, Sweden).[21] The fact that Stora Enso has sold out its forest holdings in Finland in order to focus on production has also given rise to some thoughts in SCA of ways to further rationalize its production (SCA, Sweden). In a similar vein, the prefabricated house manufacturer Älvsbyhus has put forward the prospect of moving some of its production abroad, as it considers its reasons for staying in Sweden to be cultural rather than economic. While production might become cheaper abroad, the interviewee at Älvsbyhus felt that if a factory were set up in a foreign country, the management would need to live at the location. Even if management moved, however, he was not confident that it would be possible to motivate employees and create an 'Älvsbyhus spirit' elsewhere (Älvsbyhus, Sweden).[22]

In Finland, as in Sweden, there exist similar concerns about the possible relocation of large production units, such as the Stora Enso unit at Kemijärvi. 'It would be a catastrophe if the mill closed' (Lapland Forest Centre, Finland).[23] However, as actors noted, in Finland, timber is already being imported to a large extent, in particular from regions such as northwest Russia, where the requirements concerning forest management, taxation, environmental matters and forest regeneration are different. 'It is clear that one can import timber ... cheaper where the requirements for regeneration are not at the same standard as ours, for example in Russia' (Forest Owners Union in Northern Finland).[24] This import situation was emphasized by interviewees. Some noted that the timber imports make Finland more vulnerable to any internal developments in Russia:

> *Of course Russia has plenty of forest resources, but if the forest industry takes off there, they are going to need them themselves ... that's where all the timber comes from for the mills in Eastern Finland and without the foreign wood, the industry there wouldn't be nearly that big.* (Common Forest II, Finland; cf. Common Forest I, Finland)[25]

Long-term contracts are also seen as governing import arrangements while local production is used more flexibly (Forest Owners Union in Northern Finland). This means that imports are planned, while local forestry may find it more difficult to sell timber at competitive prices. Actors considered the timber imports both as negative – in that they constituted a threat to forest owners – and as positive – in that the refinement at least took place in Finland, providing employment (Forest Owners Union in Northern Finland; Common Forest II, Finland; Common Forest I, Finland). The goal of selling wood at competitive prices is also hindered by the Finnish market situation, in which there are relatively few buyers competing for Finnish resources (Common Forest II, Finland; Common Forest I, Finland). Concerns regarding the relocation of large production units, for instance, thus centre not only on employment but also on the concomitant decrease in local timber buyers (Lapland Forest Centre, Finland). This contrasts with the situation in Sweden, where the main concern with regard to local timber is to increase its availability to the many competing local buyers; while these buyers may purchase timber from Finland, they often prefer locally produced timber on account of the lower transport costs, a factor of some importance for small-scale buyers (Sawmill Owner, Sweden).

An additional problem in both the Swedish and Finnish case study areas is the limited added value to be gained through local refinement. Limited refinement is sometimes connected to limited entrepreneurship. As one actor noted, 'The municipality of Savukoski lacks an enterprise culture, whereas in Ostrobothnia, for example, you find a business in every home. A business culture is lacking and the mentality is different' (Metsähallitus, planning).[26] A clear answer to why refinement is not more widespread was difficult to find, however, and most explanations cited a lack of funding for setting up the necessary businesses, a lack of tradition and entrepreneurship culture, and a disparity between production and the market (cf. Metsähallitus, planning). Where this last problem is concerned, one actor noted, 'Those who take the decisions know their own things, the person who decides about investments deals with customers for specialized products ... You sit with your own and think about your own' (Local Government II, Sweden).[27] In this view, limited organizational structure and interaction between actors was one reason for the low level of refinement.

3.2.4. The changing role of the state

The change in employment patterns and the loss of a local connection to the forestry resource have led to a feeling that resources have been capitalized elsewhere, far from the local community. As one actor said: 'In terms of regional economy, forestry is incredibly important because it is productive, but the jobs are elsewhere' (Metsähallitus, planning).[28] Actors agreed that forestry cannot support villages like it did before (Metsähallitus, planning; Local Government, Finland; Local Government II, Sweden; Sveaskog, Sweden).

State ownership of large production units is also seen as fostering dependence and being unreliable when it comes to remaining in the areas. For the Swedish area, one local politician noted:

> *The governing boards for ... Sveaskog have been here to have a look but when I say 'you own seventy per cent of the municipality and have some responsibility' they agree with me but you never notice any measures after they have been here ... they say that it is not us but our [daughter] companies and then they sell out these companies.* (Local Government II, Sweden)[29]

However, it may also be noted that state ownership results in at least some constancy in the areas in terms of forest investment, as state ownership of forest will provide for logging and forest production in the future as well as for some employment (Metsähallitus, planning). The forestry sector regularly calculates its logging levels in accordance with what are seen as sustainable levels of logging to ensure there will be enough timber in the future. Consequently, forest owners regularly make a commitment for over 90 years in the areas and commit themselves to long-term forest management so that they can achieve good output or sell their holdings for a good market price if they so wish.

In forestry, one can thus see that economic factors create considerable pressure for restructuring, especially due to international production networks. Interviewees described an extensive change towards internationalization during their time of employment. Even small-scale enterprises described international networks and trade as a survival factor. Similarly, local employees were concerned about the internationalization of production and the relative dependence on international markets, with trade networks – especially those including large forest companies – increasingly seen as becoming detached from the local level. On balance, the local forest industry sees itself as comparatively vulnerable in economic terms within a larger market and especially with regard to employment.

3.3. The political network influencing the forest industry

The legislative and organizational environment provides certain possibilities to deal with and cushion change. In the understanding of interviewees, there are relatively limited possibilities for support for forestry at the local and regional levels at which they act and have access to actors. This section describes the local, regional, national and international politico-legal framework as well as the actors that the local residents saw as impacting them. The state is the largest and most influential actor where legislation and access rights are concerned. In some respects, however, change may lie beyond the abilities of the state in that it is brought about by measures the state cannot influence. The state may also limit rather than only support development in the area, as described above with regard to access to wood resources in Sweden.

The relevant political context described by stakeholders is a blend of domestic and international measures. There is a mixed domestic–international context of decision-making on the environment in general and the attention to nature protection and nature reserves in particular. This issue was discussed frequently by interviewees in forestry, and was often seen as a major outside infringement of their possibilities to earn a living. Actors thus perceived themselves as having limited possibilities to exert influence at large and being in a situation where levels of governance predominantly beyond their reach impact them and determine their adaptive capacity. Interviewees also viewed these levels as guided by norms that may differ from the focus on production that pertains in much of local forest industry. The views of actors in environmental protection regarding, for instance, environmental and biodiversity goals have not been included in the study.

3.3.1. Support possibilities at the local and regional levels

Given the thin employment structure (few viable occupations) in smaller communities, forestry takes on a relatively prominent role. Local government considers itself as generally having very limited opportunities to support the industry and growth in the area due to what it perceives as its limited power, remote relation to the state and remote relation to industry. Thus, for instance, the Regional Council of Lapland in Finland noted that 'influencing is difficult' (Lapland Regional Council, Finland).[30] Correspondingly, a Swedish local government actor noted: 'Our own power is too small and this deregulation of activities means that where there is no market, there is no interest either' (Local Government II, Sweden).[31] The capacities of local government are limited to providing expertise, for example contacts with regional bodies, maintaining infrastructure, and providing measures that enhance people's image of the area and of living in the municipality. For instance, in the municipality of Älvsbyn in Sweden, the communal bath houses, ski lifts and local buses are free of charge as a result of Älvsbyn's aim to promote itself as a desirable location for commuters to larger regional centres (Local Government I, Sweden). Financing investment in the regions, from a local or regional municipal perspective, is difficult because of the aging population structure and outmigration. 'You write off a building in maybe thirty years so somewhere that perspective exists when you start; for some buildings it is perhaps fifty years' (Local Government II, Sweden).[32] These problems are especially marked in small inland communities, which are more remote and cannot gain commuters or gain from the pulling power of nearby larger towns. The support for business is thus mainly indirect and consists of providing what are limited means for the underlying infrastructure.

The principal impact of regional-level activities at the local level lies in regional planning and the implementation and monitoring of compliance with legislation. Regional-level capacities mainly involve implementation, enforcement, and regional planning of the use of land and forest (Forest Owners Union in Northern Finland). The regional level, however, cannot influence the content of regulations. Given that the regional level is the principal national administrative unit having actual connections with the local level, this also means that local complaints or comments cannot be addressed directly to a known local agent. For instance, in Sweden, the County Administrative Board noted that when it requests comments on a specific issue and details of regulation, it receives comments on much broader questions than it can deal with and does not have a channel for addressing them comprehensively (County Administration, Sweden). Similarly, the regional branch of the state forest association noted:

> *We have our role to monitor the implementation of regulation ... the laws are not always written the way that forestry or reindeer herding would like them to be written ... sometimes they have large expectations of us that we [cannot fulfil] ... but we are by far most accountable to society, Parliament and the Government.* (County Forestry Board, Sweden)[33]

The ability of regional authorities to enforce legislation is also limited by their resources for the purpose, which have been scaled down in recent years. Computerized and satellite information provides for effective surveillance over areas, but as far as the reduction of personnel goes, 'we have reached the pain threshold' (County Forestry Board, Sweden).[34] In both Sweden and Finland, EU projects have been used to fund some of the work of the branches of the regional administration in order to compensate for some of the state funding that has been withdrawn and also to add functions such as cooperation between forestry and agriculture (County Forestry Board, Sweden; Lapland Forest Centre, Finland).

3.3.2. National legislation and support

The national-level legislation and support system constitute the regulatory framework for forestry within which it adapts towards increased profitability and competitiveness. National-level legislation was mentioned both as an obstacle and a source of support. Swedish interviewees noted that forestry legislation was rather limited in comparison with the other requirements placed on forestry: 'The Forestry Act is not a problem; in fact it is quite tame in comparison with all other demands' (Common Forest, Sweden).[35] It is thus possible, for instance, to acquire a permit for logging according to legislation but to be prevented from logging by voluntary, market-driven forest certification measures regarding socially and environmentally responsible wood production that most companies have agreed to in some form. In such a situation, what may be problematic for the individual actor is that the state pays no compensation for logging being limited by environmental demands, as logging is still allowed according to law (Common Forest, Sweden).

In both Finland and Sweden, the state has limited possibilities to enforce forest regeneration legislatively, something interviewees regarded as a problem. For instance, in Sweden the regional authority can enforce replanting of a logged area only if replanting has not occurred ten years after logging: 'in such a case we lose at least eight forest years' (Älvsbyhus, Sweden; cf. Common Forest II, Finland, Common Forest I, Finland).[36] The Forest Owners Union in Northern Finland also noted that private forest owners face a problem with the law on competition, which forbids them from joining in cartels to set a minimum price and thereby putting pressure on the few wood buyers (Forest Owners Union in Northern Finland). To ameliorate this, an office with an ombudsman has been set up to assist forest owners in getting better prices (Forest Owners Union in Northern Finland). This is a problem for sellers, especially as the timber trade in Finland was previously based on a recommended price and agreements. Presently, the timber trade and price levels are based on competition in free markets, which means that cheaper Russian timber can outcompete domestic production (Common Forest II, Finland; Common Forest I, Finland). An interviewee at a common forest in Finland, however, noted that the Act on Common Forests, which has recently been amended, has increased the possibilities for forest commons to influence forest development in their areas and provide the conditions for action (Common Forest II, Finland).

Beyond legislation, many respondents discussed the support (or counteracting measures) available through taxation, economic support/subsidies and financing. Taxation was mentioned particularly frequently in Finland on account of a recent change in taxation whereby income from timber sales rather than forest holdings is taxed (Forest Owners Union in Northern Finland). This was seen as a relevant change but has resulted in some administrative and market problems during the transition phase (Forest Owners Union in Northern Finland). It was felt that the tax reform may, however, to some extent discourage logging compared with the previous system (Forest Owners Union in Northern Finland). In Sweden, taxation was less readily discussed, although an actor at a common forest in Sweden noted that one problem common forests face is taxation status: as a profit-making corporation, it is unclear if it should be taxed as private forest or as a corporation (Common Forest, Sweden).

Forestry actors in Finland also mentioned the conditions for economic support from the state. In Sweden this issue is not relevant, as forestry does not have a specific state system of economic support but is one of the few branches that is entirely self-sufficient (with the exception of support for road development and maintenance). In Finland, it is possible to gain support from the state organization Kemera if at least half of an ownership area is privately owned (Common Forest I, Finland). This provides support for forest commons. Some actors, however, argued that forestry should receive higher subsidies – in particular for transport and fuel – as its rightful level of support should be compared with that of the agricultural sector (Common Forest I, Finland; Forest Owners Union in Northern Finland; Transport Entrepreneur, Finland; Common Forest II, Finland). The importance of state support for road maintenance was noted in both countries, with a considerable consensus among actors (County Forestry Board, Sweden; Common Forest, Sweden; Norrbottens Läns Skogsägare, Sweden; Local Forestry Association, Finland; Lapland Forest Centre, Finland). Both Finland and Sweden have well-developed networks of roads, but the funding for road upkeep in general has diminished at the same time as large and heavy lorries have become more common, increasing the demands on roads (Metsähallitus, management). In this situation, there is a risk that private roads will not be maintained and that road upkeep will be prioritized in clear-felling areas (Metsähallitus, management). This might make the transport of timber in the future a problem (Lapland Forest Centre, Finland).

Actors thus noted that legislation, support systems and taxation systems have an impact on how they undertake forest management and which decisions they make. State policy on timber use over time and state regional support policy also impact local forestry, sometimes resulting in risks for the community. Especially in the Swedish case, what was seen as the state's vacillating policy was criticized sharply. For instance, the interviewees noted that the state policy in the 1970s, when the state owned a higher percentage of productive forest than today, sought to gain as high a yield as possible from the state forests. Timber from the state forests was largely sold to state-owned rather than private enterprises (Sawmill Owner, Sweden). 'They said that they would use the raw material for themselves and their own units ... we had ... to stand ... in the doorway to ask if we perhaps could buy any raw material' (Sawmill Owner, Sweden).[37] Today, on the other hand, as employment in forestry has decreased, the state has instituted measures to support local forestry employment. For instance, some 10 per cent of Sveaskog's land is made available for sale to private actors today in order to ensure them access to timber (Sveaskog, Sweden; Sawmill Owner, Sweden). However, state forestry in Sveaskog still holds a dominant position in the areas today: 'We [are still] ... at the mercy of Sveaskog and its policies ... The sawmills that remain ... have constant raw material problems ... and that has caused the number of people employed in the forest industry to drop' (Local Government II, Sweden).[38] A sawmill owner also noted – a point further underlined by other actors in a later focus group meeting – that there was no coherent state support policy for the area: while the state today supports larger private forestry ownership, state funding organs such as Almi and Norrlandsfonden do not view forestry any more favourably than private-owned banks do (Sawmill Owner, Sweden; County Forestry Board, Sweden; Stakeholder meeting, forestry, Sweden).

In Sweden, actors thus placed a major focus on the general orientation of the state and, among other things, the need for regional policy decisions in order to support the sparsely populated inland areas. However, at the time the interviews were conducted, a local government representative noted that the largest problem for the inland municipality of Arvidsjaur was not related primarily to changes in the role of the forestry industry but to whether the local army regiment, which employs many young, tax-paying people who might otherwise leave the area, would be retained (Local Government II, Sweden). The ultimate decision was that the regiment would be closed down and the jobs lost replaced with other (but still uncertain) state employment, something that provoked great worry in Arvidsjaur when the interviews were being conducted (and a demonstration involving some 4000 people in an area where the population of the central town is about 5100) (Local Government II, Sweden).

Situations such as these illustrate the declining role of the state locally compared to that in preceding decades, as its ability to invest in and fund employment in the region has become more limited. In a corresponding small community in Finland, such a situation was seen with a greater emphasis on the support structure that can be supplied by smaller actors. An interviewee at the local common forest noted that with a turnover of about 1.5 million euros per year it was the municipality's largest taxpayer. An average of over 300 euros of the association's turnover is paid to each shareholder, with several getting some 2000 or 1000 euros net in addition to their own timber sales and indirect employment effects (Common Forest II, Finland).

3.3.3. International normative changes: Environmental protection as a priority

The forestry sector sees itself as impacted by multiple factors involving regulation and policy on several levels. One of the principal problems emphasized by stakeholders in forestry in both Sweden and Finland was the external restrictions upon forestry resulting from demands for increased nature protection and concern for the environment. Actors saw this environmental concern reflected in increased environmental awareness and the inclusion of environmental goals in forestry and in state forestry bodies; in increased market demands for environmentally friendly production of wood, which manifest themselves in market-driven forest certification measures; and in more areas being set aside for environmental protection due to both external (international) demands and domestic pressure. These factors will be discussed below. On this question, local interviewees in forestry mainly focused on timber production and economic demands. External actors, as described by the local interviewees, were seen as largely focusing on strong and externally determined environmental demands and concerns.

Local interviewees in forestry generally described the heightened focus on nature protection as an attitude change reflecting broader changes in legislation and international environmental awareness. The Lapland Forest Centre, for instance, pointed out that a change in attitudes has taken place in the last ten years. The contrast is particularly striking compared to the situation in 1970s, when there were large fellings and essentially no attention paid to protection (Lapland Forest Centre, Finland). Similarly, in Sweden, as an interviewee commented: 'Earlier it was more that you thought about how to utilize the forest, it was about production more than about reserves' (Sawmill Owner, Sweden).[39] The change towards nature protection has largely taken place since the 1980s: 'These environmental matters having to do with forest nature care were not talked about until the mid-1980s' (Lapland Forest Centre, Finland; cf. County Forestry Board, Sweden).[40] And environmental matters have become much more intense since 1990:

> *It has become an entirely new division of our forestry from the bottom up, a new division, and then we had to start registering nature values ... and then came comprehensive training, we even employed an ecologist of our own ... things are very different now ... then we had to certify, as well.* (SCA, Sweden)[41]

3.3.3.1. Nature reserves

The most apparent effects of the increased focus on nature protection are perhaps seen in ongoing protection processes. Following Swedish and Finnish accession to the EU in 1995, both countries have had to adopt – rather quickly – EU regulation, including Natura 2000, the EU network for protected areas, which adds to domestic regulation. Metsähallitus has ongoing processes of nature protection, to a large extent related to the selection of Natura 2000 areas. This external influence on nature protection was noted by some actors: 'If a Natura area is involved, the decisions are made elsewhere' (Lapland Regional Council, Finland).[42] The Finnish Association for Nature Conservation and the WWF in Finland have proposed additional protection areas (Metsähallitus, management). In Sweden, the County Board has recently made an inventory of state forest lands for protection in response to a request by the Swedish Environmental Protection Agency as a follow-up to a domestic assessment of reserves and the Natura 2000 process. The County Council Administration in the Swedish case views Natura as the focus of the conflicts: 'Right now a lot [more forest] is set aside to develop nature protection reserves ... I think that right now forestry feels very pressured by this Natura 2000 development' (County Administration, Sweden; cf. SCA, Sweden).[43] Given that a large percentage of the inland area around Arvidsjaur and Arjeplog is state land and thus falls within this inventory, there is a fear that the inventory and subsequent decisions may exclude additional forest areas from production (SCA, Sweden). The northern inland area is where most old-growth forest can be found and where most of the large nature reserves have already been developed (SCA, Sweden). Actors also worry about state policy goals of protecting some 900,000 hectares by 2010, some 350,000 of which the County Board has suggested should be in the county of Norrbotten: 'Then we would account for a third of the protection in this country and then you understand why forestry wonders how this all is going to play out' (County Forestry Board, Sweden).[44]

Common forest representatives in both countries also similarly noted and expressed concerns about protection (Common Forest, Sweden; Common Forest I, Finland; Common Forest II, Finland). An interviewee in Finland noted that in other common forests 'they have had problems with protection and the compensation received for the land they've given up is inadequate' (Common Forest I, Finland).[45] Although the focus in environmental protection today is on state forests, private actors worry that it may come to affect them, too: 'There is this fanatical protection which fortunately has not hit … the common forest yet' (Common Forest I, Finland).[46]

The greatest worry generally among actors was that environmental protection would further impact production, especially since local entrepreneurs need to provide local raw material at marketable prices to survive (Sweden) and local timber production has to be able to compete with imported timber (Finland) (cf. Lapland Regional Council, Finland; Sawmill Owner, Sweden). The worry in Sweden was that increased environmental protection would limit the amount of local timber available on the market and make large actors such as SCA less inclined to invest in the area (County Forestry Board, Sweden; SCA, Sweden). Local interviewees thus emphasized what they saw as an excessive level of protection in relation to the negative effects of protection on living conditions and employment: 'It impacts the local population' (Forestry Machine Driver, Sweden; cf. Local Government II, Sweden; Metsähallitus, planning).[47] In Finland, interviewees noted that the effects on the timber supply of the establishment of Urho Kekkonen National Park (some 800 square kilometres of protected area) could be seen at the municipal level in decreased tax revenues, among other things (Local Government, Finland; Transport Entrepreneur, Finland). 'That is why our municipality has been in dire straits – the economic basis of its income has changed' (Local Government, Finland).[48]

The main problems perceived among actors with regard to environmental protection were thus economic; and it was the level rather than the principle of nature protection that concerned them: 'Every cubic metre of forest that is logged has a GNP value of about [SEK] two and a half thousand, so it has consequences, economic consequences, if you "remove" forest land from production' (County Forestry Board, Sweden).[49] Similarly, another actor noted: 'I am the first to go along with environmental protection but it shouldn't always involve sacrifice; you need to increase growth, because if … you cut flow of resources into the region you get no immigration into the area and no growth' (Sawmill Owner, Sweden; cf. Metsähallitus, planning; SCA, Sweden).[50] Interviewees thus perceived the levels of environmental protection as limiting the available land for and income from forestry. For instance:

> *If they start setting aside all the nature reserves they have planned to in this area, then it is like sitting in your own larder and not being allowed to eat … one has to be well aware that if you are to run an industrial operation inland in sawmilling, then you need timber. Otherwise … we [as entrepreneurs] remove our capital and sit wiggling our toes, but there is no employment.* (Sawmill Owner, Sweden)[51]

The different conceptions of forestry and environmental protection were also seen by interviewees as potentially harming the sector. Many actors especially noted the risk that northern areas would start to be perceived as being designated for recreational and environmental protection rather than forestry. For instance, the Lapland Forest Centre interviewee observed: 'The office of Stora Enso is in London … [They] may well think that, hold on, Lapland is a recreational area for Europeans so you probably can't cut down any trees. The [mentality] might be something like this' (Lapland Forest Centre, Finland).[52] In northernmost Finland (outside the case study area) conflicts have arisen about the use of old forests for reindeer herding, among other things. The Forest Owners Union in Northern Finland mentioned that this conflict over environmental matters was not good for the image of northern forestry; it can become a risk for companies even where the conflict mainly involves state forest (Forest Owners Union in Northern Finland).

Interviewees recognized, however, that today's environmental protection levels constitute an improvement over and sometimes a necessary change from the production-oriented forestry of the 1970s (Forestry Machine Driver, Sweden; cf. Metsähallitus, planning). Actors in forestry have also begun to emphasize, and to some degree adapt to, multiple economic uses of forests, of which forestry is but one. For instance, the Swedish National Board of Forestry underscored 'this social dimension'[53] 'beyond the traditional Forestry Act' (County Forestry Board, Sweden).[54] The Forest Owners Union in Northern Finland similarly noted:

> *Our thought is that this kind of multipurpose entrepreneurial activity must be accepted here, it's not only agriculture, only reindeer herding, only tourism or only forestry … [if] one wants to maintain entrepreneurship in sparsely populated areas, we have to recognize that entrepreneurs get their income from many different sources depending on the season.* (Forest Owners Union in Northern Finland)[55]

Similarly, one of the Finnish common forests representatives said that they were trying to branch out by buying more land that can be added to the common forest and by developing additional uses of land such as tourism routes or renting places to tourism entrepreneurs to 'keep other options open if the income from forestry decreases' (Common Forest II, Finland; cf. Common Forest I, Finland).[56]

3.3.3.2. Certification: Socioeconomic and political globalization?

Another issue closely related to environmental protection is forest certification, which was frequently discussed by actors. Forest certification, developed particularly in the last ten years, is a market-driven initiative entitling a company to label wood products as certified if they meet certain environmental and social requirements. The criteria for certification are set up through the development of international and national standards and monitored by independent auditors. Forest Stewardship Council (FSC) certification has the more demanding set of forest certification criteria, advocated by environmental organizations, and is the most common form of company certification in Sweden. The Programme for the Endorsement of Forest Certification (PEFC, previously known as Pan-European Forest Certification) is a somewhat less demanding standard, which is the prevailing one in Finland and the one adopted primarily by private forestry in Sweden. The number and intensity of interviewees' statements on certification reflected this difference between the standards. In Finland certification was discussed less, being mentioned only by a few actors, while in Sweden it was discussed on a par with environmental protection.

In Finland, the comments on certification were less specific, possibly as the form of certification in the country (PEFC) is generally considered less demanding. It was noted that certification is even of popular interest. The Lapland Forest Centre, for example, noted that certification is important to some degree for tourism, as the tourist centres get questions about the forests around them and whether they are sustainably managed (Lapland Forest Centre, Finland). The salient consideration with regard to certification in the Finnish material was the import of Russian timber: 'Russian timber is not certified and they tell us that every day they have customers asking for the origin of the timber and from what kind of forests it is cut' (Lapland Forest Centre, Finland).[57]

In Sweden, on the other hand, interviewees spontaneously discussed certification to a very large extent, bringing up the problems, possibilities and constraints that they felt certification had caused. Most companies are already certified or are trying to get certification. For instance, one actor said: 'We think a lot about how far we are going to go with this … there is a rather big market demand from Britain above all' (Sawmill Owner, Sweden).[58] One of the common forest organizations interviewed was among the few companies that had not yet been certified, noting that if they were, they would need to at least double the size of their voluntary protection areas. This is because the FSC certification scheme demands that at least 5 per cent of forest holdings be protected but leaves the upper limit on protection dependent upon the protection value of the forest (Common Forest, Sweden). This requirement especially impacts areas which in terms of certification are considered to be beyond productive age and to include protection values, but where Swedish legislation allows for logging: 'Virtually all timber buying companies are certified, so they make stricter demands than the Forestry Act does'

(Common Forest, Sweden).[59] As a result, some actors may end up with timber they cannot sell yet none of the compensation they would have received if logging had been prohibited according to the Forestry Act. 'It becomes a Catch 22: you get permission to log but you cannot get buyers ... Since the National Board of Forestry permits logging ... the National Board of Forestry does not pay compensation' (Common Forest, Sweden; cf. Norrbottens Läns Skogsägare, Sweden).[60] Even actors who are not certified may be prevented from selling timber from such areas: 'If the media makes a big deal of the fact that old forest is being cut ... forest companies do not dare buy that timber' (Common Forest, Sweden).[61]

Given the structure of the forest in the regions, however, some of the interviewees argued – in a similar vein to arguments concerning nature reserves – that certification was symptomatic of an external view of northern areas and not locally adapted. The FSC was perceived as 'developed for forestry where only productive forest exists ... and where there do not already exist so many nature reserves. It is not at all suited for the inland areas here' (Common Forest, Sweden; cf. Sveaskog, Sweden).[62] Additionally, from an outside perspective, while 160 years may be a very old forest elsewhere, in these areas 'if your full growth cycle is 130 years, 160 years is not a very old forest' (Common Forest, Sweden).[63] On this account, 'certification is a good thing, but it has not been developed such that it suits our forest areas' (Common Forest, Sweden; cf. Forestry and Reindeer Herding Administration, Sweden).[64]

A further risk perceived by some interviewees was that, like environmental protection in general, certification may contribute to negative economic and population development in inland areas. The FSC aims to ensure production as well as to preserve environmental and social values. However, one interviewee felt that 'it is only the environment that counts ... the other two legs [socially and economically sustainable forestry] are only support legs' (Common Forest, Sweden).[65] While social values in the FSC framework focus to a large extent on indigenous peoples' rights, the same interviewee also noted that social values should include values for the socioeconomic area where logging takes place, such as employment (Common Forest, Sweden). A Swedish common forest interviewee noted that:

> *... social values are that you can continue living on your holding, that you can make some money from the forest because there are forest entrepreneurs and drivers, and together there are enough of you that you can keep the grocery store in the village going, and the school and so on. ... These are social values, too. If you set aside most of the forest land, the economic vitality of the village stops and it becomes depopulated.* (Common Forest, Sweden)[66]

Some actors in reindeer herding (a largely indigenous occupation), however, note that FSC certification has improved the communication between forestry and reindeer herding, as it provides for consideration for indigenous practices (Forestry and Reindeer Herding Administration, Sweden). In some areas, state regulation has required forestry interests to consult with reindeer herders regarding planned logging. Today, these requirements have been extended by FSC certification to all certified actors within the reindeer herding area (Forestry and Reindeer Herding Administration, Sweden). However, neither procedure imposes a requirement that forestry companies come to an agreement with reindeer herding representatives before logging. Conflicts between reindeer herding and forestry thus exist regarding regulatory frameworks for multi-use forests (including both legislation and certification), but have a larger impact on reindeer herding than on forestry.

On balance, despite concerns over constraints to production, interviewees in both Sweden and Finland shared the view that certification is a relatively well-accepted, even self-evident, process compared with environmental protection in the form of nature reserves. This view may reflect the fact that certification is a market demand and thereby directly impacts opportunities to sell timber and wood products to export partners. Environmental protection by the state or determined by the EU (for example Natura 2000), on the other hand, is perceived as a non-market-related external regulation that diminishes the amount of timber available for production without adding positive market benefits. An interviewee at Metsähallitus noted, for instance, that certification benefits the image of forestry and that without certification it could become difficult to export timber (Metsähallitus, planning). Although the pros and cons of certification have been debated a great deal, it is now largely seen as accepted (Lapland Forest Centre, Finland): 'It is the customer demands that have made themselves felt throughout Europe' (SCA, Sweden).[67] An interviewee at the common forest said, 'I believe that in time there will be some sort of certification for most forestry. It is a demand from buyers so they can say to their customers – for example these large building supply firms – that this [wood] comes from forestry conducted in a special way' (Common Forest, Sweden).[68] Certification as a market demand by customers is thus to a large extent seen as unavoidable; state and EU regulation, however, are seen as both external to the market system and as something that actors may have a right to lodge a complaint about. One actor even went as far as to claim that nature reserves and state actions to some degree duplicate the dictates of the market, thereby increasing the demands for environmental protection many times over: 'We have adopted … certification, we have not noticed that the market has any other plans … but that is not enough: more forest is still to be set aside' (SCA, Sweden).[69]

3.3.4. The governance network as perceived by local forestry

To summarize, the governance framework as described above includes actors and frameworks that are both economic and political: the EU; state administration and national legislation; state support organizations, such as Kemera (Finland); state environmental protection and environmental protection agencies, such as the Swedish Environmental Protection Agency; environmental protection organizations, such as the WWF and the Finnish Association for Nature Conservation; international timber buyers, such as SCA, Stora Enso and the international market; the PEFC and FSC certification schemes; state financial institutions, such as Almi and Norrlandsfonden (Sweden); banks; and public opinion. Actors made relatively little reference to each other, with the exception of well-known state-level actors (Sveaskog, Metsähallitus) and large companies (such as SCA). In general in the study, the only actors who interacted were those who either have a regulatory requirement to consult each other (for instance, forest owners, who are required by state regulation and/or FSC certification to consult representatives of reindeer herding) or who need to directly interact otherwise regarding production (such as actors who trade in timber or are parties in the same buyer–seller networks, as is often the case with the national organizations or large companies). The contacts among local and regional-level interviewees, however, were rather limited, as were vertical links between actors. An exception here was collaboration in land-use planning and consultation, especially with regard to multiple uses of forest. Thus, for instance, a common forest interviewee noted: 'We collaborate continuously with everybody ... reindeer herders, the municipality, farmers, hunters, the Forest and Park Service' (Common Forest II, Finland; cf. Common Forest I, Finland).[70]

All in all, interviewees described very limited interaction between themselves and political, administrative and sectoral actors, even in the same area. For example, a sawmill owner noted that he did not have any regular contact with local politicians even in his small municipality. He felt that this limited contact restricted politicians' knowledge of the requirements for forestry. As he said: 'Most often when you talk about the future of the municipalities you talk about getting an education ... you do not talk in industry terms; no one mentions how important it is that there remains development in the small and middle-sized corporations in the forestry business' (Sawmill Owner, Sweden).[71] A local forestry association in Finland also noted that it had little contact with the municipal government even though they were in the same small local area (Local Forestry Association, Finland). A representative of local government also felt that there was a lack of awareness of the importance of forestry in the municipality (Local Government, Finland). Similarly, regulation administrators noted this limited knowledge, sensing that there was a distance between them and decision-makers at the governmental level: 'There is a reality that they have no idea about – that [certain] problems can exist' (County Forestry Board, Sweden).[72]

Horizontal integration between actors in the same organizations or direct fields of interest was also limited. To some extent, this was a result of the different administrative units being rather dispersed. One interviewee expressed this in the statement 'forestry is sparsely populated' (County Forestry Board, Sweden).[73] These 'thin' structures are to some degree made possible by modern technology, but also make it more difficult for those unaware of the network organization to have access to decision-makers within it (for instance because they never meet them). 'So very much has happened in the organizations ... [today we are] in reality separate individuals working on [different questions]' (Forestry and Reindeer Herding Administration, Sweden).[74] On the other hand, for administrators within the same organization who know each other, the thin structures of decision-making can also be a benefit: 'We know who to talk to' (County Forestry Board, Sweden).[75] At the same time, integration may also be difficult locally within an organization, possibly because of limited time and interest, or because of changing ownership structures in forestry. An interviewee at a Finnish common forest noted that they have some 1500 shareholders, of whom only the 50 largest attend the meetings (Common Forest II, Finland, Common Forest I, Finland).

There was also little contact horizontally between competitors in the areas, even though they know of each other's activities (Sawmill Owner, Sweden). The limited interface which an individual has is especially pronounced in larger corporations: there is little contact with those deciding on one's life situation and continued employment, and decision-making units are far away: 'The distances are too long ... too long between these places' (Forestry Machine Driver, Sweden).[76] The same actor noted that, especially for the individual employee, shortcomings in the communication within organizations become a liability during times of downsizing, as the company's concern for the local area in the long term may be limited (Forestry Machine Driver, Sweden; Swedish Wood and Tree Trade Union). In general, then, the actors interviewed did not interact much: for instance, Älvsbyhus had only some limited cooperation with one sawmill through the timber trade, but otherwise interacted mainly with its supplier, Sveaskog (Älvsbyhus, Sweden).

The interviewees' networks indicate that even within a small local community contacts cannot be assumed – and the contacts between stakeholders may, where they exist, be limited to direct economic interaction or trade. Interviewees noted, however, at the same time, that many different groups have opinions about forestry – sometimes appearing in the media – that may impact the production values of forestry: 'The forest is like a commons ... if you as a private person won two million in the lottery or something and put it in stock, you would get to manage it as you wish ... if you buy a tract of forest for two million, the public and environmental groups and everybody have opinions on how you should manage your capital' (Common Forest, Sweden).[77] Similarly, an interviewee associated with a common forest in Finland noted that it is unclear to many what a common forest is and that many perhaps therefore see it as common or state property: 'There are a lot of people in the province of Lapland and all over Finland [who] haven't got the slightest idea what a common forest is ... it is an unclear idea to many people' (Common Forest I, Finland).[78]

On balance, interviewees found the interaction in the forestry sector limited (with the possible exception of forest-use planning), and perceived differences in orientation between themselves and the international or larger-scale actors they viewed as impacting their own practices. This multi-level governance network

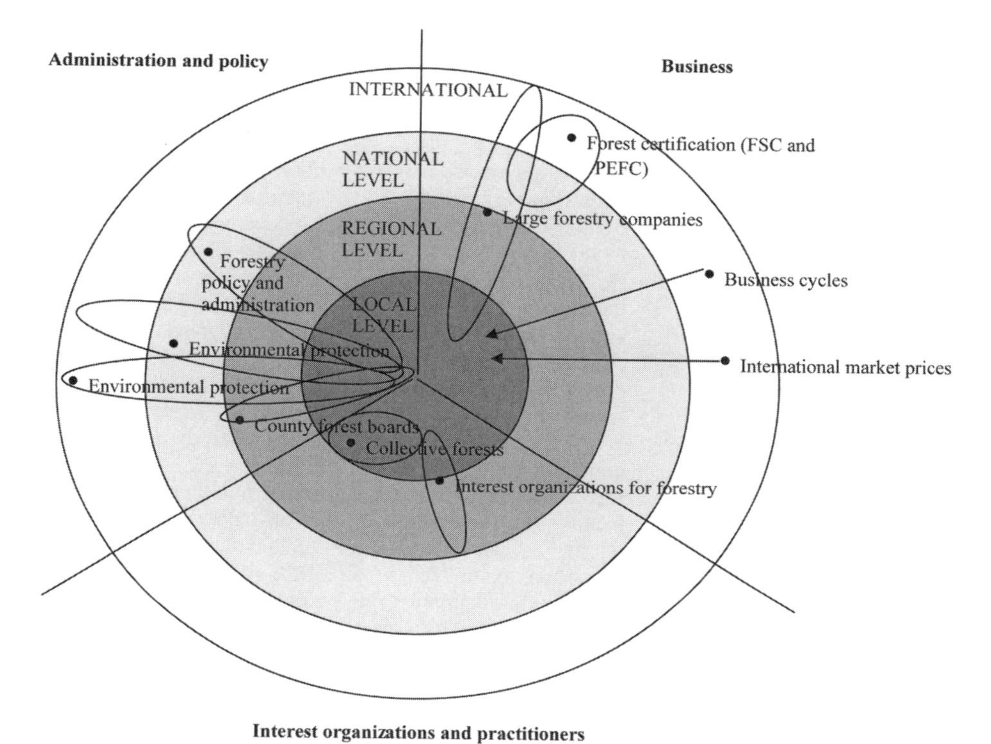

Figure 3.1 *Multi-level governance of forestry*

The diagram is divided into three sectors corresponding to the main groups of actors (administration and policy, business, and interest organizations and practitioners). Four levels of organization are distinguished: local, regional, national and international. The figure shows the principal organizations active in each sector, with dots indicating the level at which each nominally operates. Ellipses reflect the scope of an organization's influence as perceived by interviewees and arrows indicate large-scale influence via the market or regulation.

is illustrated in Figure 3.1, which indicates the actors in forestry that the interviewees themselves discussed.

Figure 3.1 is divided into administrative, industrial and civil organizations at the local, regional, national and international levels. It illustrates that actors mainly in administration and policy, and in general actors at the national level, dominate the organization of the forestry sector. The figure also illustrates that very few of the actors discussed by interviewees are organized locally. One inference that can be drawn from the figure is that adaptive measures, even for local problems, would need to span several levels of organization – including actors with whom interviewees have no clear link. Interviewees also described the increasing importance of norms and regulation at the international level, such as environmental norms emanating from the EU, national protection measures and voluntary market certification systems.

Table 3.1 *Main perceived changes, adaptations and limits on adaptation in forestry*

Changes in the governance framework (economic and political changes)	Adaptations to perceived changes	Limitations to adaptive capacity
• Rationalization and decreased employment • Shift from employed forest workers to hired contractors • Increased competition • Technological development • Increased importance of multi-use and environmental protection • Structural changes in forest ownership and interest • Increased discussion of conflicts and protection/ concern; forestry perceived as distanced from people's perceptions	• Companies moving or threatening to move; increased emphasis on technology • Increased scale benefits to larger units (for e.g. large-scale contractors and sawmills) and more limited connection to communities • Awareness of production conditions in other countries and limitations in resource access; focus on export niches with higher returns • Emphasis on need for loans for reinvestment • Multiple uses and tourism • Alternative resource-sourcing routes	• Lack of entrepreneurship culture for refinement • Limitations in compatibility between forestry legislation and certification (limiting compensation); certification requirements seen as being of limited suitability for the region • Limited legislative enforcement of forest regeneration or encouragement of private smallholder logging • Limited support for road maintenance; limited ability of different levels of decision-making and management to guarantee infrastructure at large in the areas • Constraints to forestry production due to environmental protection aims

3.4. Climate and climate change

The governance network described and established values in forestry provide the context in which actions on climate would need to be communicated and adopted. The possibilities for such actions are, however, limited both by the characteristics of the network and the rather compelling processes of change in the areas, which limit interaction on issues perceived to be long-term in nature. Climate change must thus be seen as much in the context of socioeconomic and administrative choices as in relation to direct climate effects.

The sections below describe the impacts interviewees anticipated for the different seasons, corresponding to the way in which stakeholders described change. The results show that climate change may have both positive and negative impacts. It seems that negative impacts will mainly take the form of adjustments to changes, and additional costs from such things as maintaining and building roads, to more long-term and potential changes, such as changing customer structures in response to changes in wood quality. The positive impacts cited included increased growth rates. Climate change thus will affect production conditions to a large extent in the short term, one example being transport. Changes in the arrival of autumn and spring – typically transition times between more stable states – will have similar impacts on actors, given that these seasons involve changes across the freezing point. Predictability and the importance of meeting production schedules for consumers were emphasized; changes that alter schedules, such as early thawing, will impact production (Transport Entrepreneur, Finland; Sawmill Owner, Sweden).

Actors indicated some urgency related to the climate in that certain potential weather changes have been observed in recent years. While these few years may be atypical rather than indicative of a trend, they prompted interviewees to cite ongoing and not only possible adaptations or directions of adaptation to change. It is difficult to distinguish a clear trend, however: interviewees noted that relatively hot as well as relatively cold years have occurred (since the 1950s, as one interviewee recalled) (cf. Common Forest I, Finland).

3.4.1. Climate impacts on forestry in autumn and winter

In autumn, one of the most important things for forestry is a relatively quick transition to winter and stable frozen ground. Forestry is dependent upon the ground staying frozen in winter, as this allows access to areas where the roads otherwise could not bear logging lorries for transport or forestry machines for logging. 'The best thing of all when it gets cold is that the ground and roadbeds freeze solid before it snows. It is not slippery, the roads hold up and if you fill up a ditch it freezes, and so on. That's the best situation of all' (Transport Entrepreneur, Finland; cf. Common Forest, Sweden).[79] If roads and the ground do not freeze, machines sink and cannot access areas for logging and transport (Transport Entrepreneur, Finland). Thaws may also result in wetter snow, which in turn affects wood quality by weighing down tree crowns at the edges of plantations, where there are no surrounding trees to prop them up (Common Forest, Sweden).

Thus, cold weather should continue when it has started, which may be a problem given the prospect of prolonged warmer periods or periods in which the temperature fluctuates around the freezing point. 'If the autumn is cold and it looks like winter is coming with cold, frozen conditions and everything, we may go onto weaker ground too early; but this backfires if the frost melts since we then have to move on to stronger ground' (Sveaskog, Sweden).[80] Changes in weather that affect the load-bearing capacity of roads are thus the most problematic for forestry in autumn.

It is thus ideal in winter if the weather remains stable and safely below the freezing point, so that no thawing will occur that may impact load-bearing capacity: 'In winter, 10–20 degrees below [zero Celsius] is ideal' (Transport Entrepreneur, Finland).[81] This temperature is ideal in that it is not too close to freezing, keeping the roads solid and avoiding slippery conditions due to thawing

(Transport Entrepreneur, Finland). If the weather were to get warmer, 'if you consider transportation, it is bad surely … first of all, you're basically guaranteed to get more slippery conditions even on the highways' (Transport Entrepreneur, Finland).[82] A temperature of –10 to –20°C is also preferable because it is not too cold for hydraulic equipment (Transport Entrepreneur, Finland); forest work may otherwise be stopped by very severe cold of –40°C or so (Metsähallitus, management; Local Government, Finland; Transport Entrepreneur, Finland). It is also ideal if there is not too much snow that needs to be ploughed off roads and tracks, which increases harvest costs (Metsähallitus, management). These costs are due to the fact that deeper snow is harder to work in and thus lowers output and entails a higher risk of damage to machines. Moreover, contractors demand higher compensation for work in deeper snow (Local Forestry Association, Finland; Metsähallitus, management).

Some interviewees observed that changes such as the above – resembling those that could result from climate change – have been taking place in the last few warm years. As one noted, 'The last five or six years we have had almost no snow before Christmas and then it has thawed after Christmas and the ground has almost become bare again' (Sveaskog, Sweden).[83] Similarly, 'The winters have been milder than they have usually been … sometimes we have practically not had a white Christmas here either and that has always been common' (Norrbottens Läns Skogsägare, Sweden).[84] While some of these recent impacts are relatively clear – such as a later winter and more winter thaws – others are more uncertain, such as whether the amount of snowfall in winter has increased or decreased. A Finnish interviewee observed, 'This warming seems to make the winter shorter but it does not decrease the amount of snow' (Lapland Regional Council, Finland).[85] A Swedish interviewee, on the other hand, observed that the climate and snow levels have changed in the direction of generally warmer weather and less snow. This has had some benefits for forestry but some negative impacts on forestry workers: 'It has become greyer, more boring in many ways [to work]' (Forestry Machine Driver, Sweden; cf. Sveaskog, Sweden).[86] For forestry, 'the conditions are pretty much optimal in the forest now, the ground is frozen solid, hard as a rock and there is little snow [and it is] easy to drive on winter roads' (Forestry Machine Driver, Sweden).[87] On the other hand, one interviewee observed that not only the warmer weather but also wind conditions have affected snow levels: 'As a rule this time of year has become warmer, there has been more rain than before … and the wind has started to blow from the west … the falling snow lands in the mountain regions' (Forestry Machine Driver, Sweden).[88]

To date, these changes have not caused major problems but have increased the costs of production:

> *Today we do not have such strong thaws that we have to reschedule logging. But on the other hand we do have costs on the roadside with road graders and sanding and such on ice on the roads so that lorries can*

> *make it through. There is some extra expense with grading and gravelling and such, but the situation is not so bad that you have to reschedule logging.* (Common Forest, Sweden)[89]

In fact, less snow may be an advantage (Sveaskog, Sweden). As effects are currently not so marked, the warmer winters with less snow in some areas have been beneficial, especially as thaws and the impacts on accessibility have not been too extensive.

The main focus for forestry where climate and weather conditions are concerned is thus accessibility to logging sites. If the winter season becomes shorter because of climate change, 'then it will bring problems ... in the winter it is possible to use winter roads which involve much much cheaper costs' (Metsähallitus, management).[90] The additional costs of using summer roads or using winter roads for a shorter period depend on the circumstances and need to be calculated 'pretty much on a case-by-case basis; you calculate the quantity of timber involved and how far it is from a normal summer road. You have to sit down and work out the costs to see if it is worth waiting and getting the wood in the winter' (Metsähallitus, management).[91] Additionally, depending on ground quality, problems may be larger or smaller: heavy machines sink most easily in fine-grained soil, while much of the ground in the case study areas is coarse-grained (Common Forest, Sweden). In this respect, the areas may have a benefit in alleviating moderate impacts of change. The extent of the road network, however, remains a limiting factor.

3.4.2. Spring and summer

Spring, like autumn, is a transitional period, with similar impacts on accessibility due to changes in the bearing capacity of forest ground and roads: 'It is in spring and late autumn that you have problems with the carrying capacity of the ground and the roads and then you have a limited supply of timber available' (Common Forest, Sweden).[92] As for autumn, the time at which the season arrives has changed: spring has started earlier, especially in the few recent warm years, with the beginning of the season being somewhat less predictable. Easter is traditionally a turning point for melting, at least in the Swedish area, and has become the time when logging needs to move out from winter road areas, which will start thawing and no longer be passable: 'If you are out on a winter road then at Easter you have to be out of there ... [that is a] golden rule' (Sveaskog, Sweden; Common Forest, Sweden).[93] However, Easter may not remain fixed as the time when the shift occurs: 'Last year it started thawing long before Easter ... so there was panic at some places' (Common Forest, Sweden).[94] The same interviewee also noted that it was a problem that there were no cold nights after thawing had begun, during which timber is usually transported during a regular year if some of it still needs to be brought out after Easter (Common Forest, Sweden). Not only the onset of thawing but also the temperature during the day and night thus play a role for transport around this time of year. In terms of losses to industry, a period of about a month in the spring is seen as the most risky; industry is not equally affected by wet weather in summer, for instance, given that the demand for industrial timber decreases with decreases in production during the vacation period (Common Forest, Sweden).

The problems in predicting accessibility are a concern especially given the short delivery times to industry: 'Since we need to get the timber to them because there are almost no storage depots anymore, sawmills have no storage facilities and the forest industry at large has none either, there is a week between logging and the wood being sawn' (Common Forest, Sweden).[95] Problems such as these may add up to substantial costs for industry: 'If we have problems with the weather then all other small suppliers have problems with the weather and together we incur a large loss' (Sveaskog, Sweden).[96] For industry, a consequence in the extreme case might be demands to lower or modify orders, which would limit or prevent reinvestment (SCA, Sweden; Sawmill Owner, Sweden).

Spring involves adaptations to changed road accessibility similar to those in the autumn. If the weather changes when logging has already taken place, one has to 'spread gravel on the road … at high costs … and is forced to take the logs out before the opportunity is lost altogether' (Common Forest, Sweden).[97] Alternatively, the situation can be prevented by building better roads; this is, however, a costly option (Common Forest, Sweden). The capacity to deal with these kinds of problems is dependent on the actor involved: '[Larger companies] have larger road networks … so they have somewhat better possibilities to prepare in advance. … A small forest owner is often dependent on the neighbour's road too' (Common Forest, Sweden).[98] Climate change could exacerbate conflicts depending on how much extra money is invested in extending the road network for logging:

> *If you spend money on a more expensive road to an area in order to be able to log it when you wish, you do not want to forgo this if the Saami village has a different interest … you can take each other's livelihoods into consideration but you have a certain limit and where it starts to cost too much, you worry about your own interests instead.* (Common Forest, Sweden)[99]

There are also problems beyond accessibility with possible shifts in the thaw and the nature of spring. Interviewees mentioned that both slow and quick melting could be problematic. In some instances, the slow melting of snow – for example as a result of higher precipitation in winter – might limit growth: 'Places with sleet and such – that is a problem of course since it impacts, decreases, growth' (Sveaskog, Sweden).[100] It was also pointed out that quick spring melting may worsen floods (Lapland Regional Council, Finland), although this may also partly be a result of direct human actions such as the digging of drainage ditches by the forestry industry (Lapland Regional Council, Finland). A long, wet spring may also damage tree roots (Transport Entrepreneur, Finland).

With regard to summer, interviewees noted that warmer summers would be to some degree beneficial, as they result in a better seed crop and higher growth rates (Metsähallitus, management; Forest Owners Union in Northern Finland; Lapland Regional Council, Finland). 'You get more frequent seed years and regeneration of forests becomes much easier' (Metsähallitus, planning).[101] Higher temperatures in particular will improve growth: 'The heat sum is exactly what restricts growth; in other words [growth] requires a certain amount of heat, the average annual temperature' (Local Forestry Association, Finland).[102] 'If you get one added day to the 60 days [of summer], then you get more than 1 per cent. If you get 6 days more then the forest grows 10 per cent more' (Common Forest II, Finland).[103] Warmer weather would thus promote forest growth. However, this higher productivity would also result in a need for improved forest management and, for instance, more frequent thinning operations. This could be a potential problem, as thinning is already limited because it is a labour-intensive measure and employment in forestry has been rationalized (Common Forest II, Finland; Common Forest I, Finland).

Interviewees also noted that changes in the growth rate, although beneficial in providing for increased growth, might affect wood quality: 'We [would] get timber more quickly but the quality would decrease ... the northern pine ... has been a large market advantage, it has been northern pine that has sold well' (Sveaskog, Sweden).[104] If forest growth increases but the quality deteriorates because forest is less slow-growing, companies may have to adapt by adjusting prices and even approaching different buyers – even though timber from these regions would still be among some of the most slow-growing on the market: 'Perhaps you cannot sell dense pine in the same way, you end up having to compete with timber from central Sweden [and selling to] Greece ... where such wood is valued. Perhaps you have to adjust the price' (Sawmill Owner, Sweden).[105]

Interviewees drew particular attention to possible production benefits. Forest is currently harvested at the age of 110 years, but 'if it reaches a certain sturdiness it may be possible to harvest it at 90 years' (Local Forestry Association, Finland).[106] In the Swedish case study area, forest is already harvested at 90 years of age – to some extent as a result of forest companies lobbying to lower the minimum logging age. This results in relatively thinner tree trunks than logging at a higher age, a trend mentioned as a problem by some interviewees (Älvsbyhus, Sweden). Stocks are dwindling in general for reasons unrelated to the climate, since certification prevents forest being cut once it is deemed to be of an age when protection values can be expected. These impacts will interact with those caused by climate change. With increased warming, however, the cycles of forest renewal might even become some 70 to 80 rather than 90 years, providing relatively thick trunks in a short time: 'If it warms up even more, it may be that the turnover time is even better, faster, maybe even 70 years' (Common Forest II, Finland).[107]

Modest warming could thus be beneficial for the region (Metsähallitus, management).

Actors emphasized, however, that it would be difficult to know if any other impacts might offset those of warming: 'Warming can be a good thing in the short term. I don't know what other effects it could have on the environment if trees start to grow faster, and are there other impacts ... which will negate that growth' (Metsähallitus, planning).[108] Some of the added impacts of warming could be precipitation changes and pests. While summer heat has not caused problems (Metsähallitus, management), dry conditions could become problematic (Common Forest, Sweden; Norrbottens Läns Skogsägare, Sweden). If, on the other hand, precipitation increases, this would require, for example, increased sanding during the transition periods in autumn and spring: '[Increased precipitation] would have a decisive effect on the state of the roads. It would of course become a question of costs' (Metsähallitus, planning).[109] In autumn and spring, increased precipitation could also result in accessibility problems, which can impact the quality of already cut timber waiting to be picked up. Such effects on timber are potentially largest in summer, when insects and pests abound: 'When we have ... several thousand cubic metres at the end of some road and we have to wait a couple of days ... until the road has dried to get the timber ... then [we] cannot sell it because there is ... blue stain and insect damage and such' (Sveaskog, Sweden).[110]

Warm summers, especially in combination with damp weather, may also increase pests (Transport Entrepreneur, Finland). Many interviewees connected the possibility of warming with the possibility of more moisture. For instance, an interviewee at Metsähallitus noted that especially if heat were combined with moisture, there would be a risk that 'fungal diseases would increase, reducing quality' (Metsähallitus, planning).[111] Some years ago, parts of the Swedish case study area had problems with pests: 'They said it was because the year before was a wet one, and there was a real [warm] summer the year after and then [the moths] spread out over a large area ... then it could be because of something, environmental changes' (Sveaskog, Sweden).[112] There have also been problems – mainly local – with the European pine sawfly and pine weevil (Common Forest, Sweden). A large increase in pests would require adaptations that for the most part are unknown in the areas today. On the whole, pests are 'something we have been spared from here ... since we cannot use any chemicals today in Sweden I do not know what we would do [if pests were to increase] ... The question is how much warmer it would have to be in order for that [to happen]' (Norrbottens Läns Skogsägare, Sweden).[113]

Another, longer-term impact of climate change might be its benefit for spruce vis-à-vis pine. This would be reflected as spruce taking over pine lands and growing on areas that become suitable for forest (even if soil quality might at present limit such a spread). If access to logging sites or wood quality changed, adjustments would need to be made to logging and production; for instance, sawmills would have to equip themselves to saw more spruce (such adjustments could take place in step with investments in new technology, thereby limiting costs directly attributable to the change in the type of timber sawn) (SCA, Sweden). This scenario illustrates the interaction between climate and economic factors. Any climatic benefit for spruce at the expense of pine would also need to be taken into consideration when planting, as forests today are planted rather than naturally regenerated (Sveaskog, Sweden). One way to prepare for possible forthcoming changes that would favour certain growing conditions could be to select the stock to be planted in accordance with the projected conditions: 'If you plant and saw you need to have good seedling and seed stock so that you can sort of adapt to this. And then it would be important to conduct research so that we would be able to offer our forest owners the right material for planting' (Norrbottens Läns Skogsägare, Sweden).[114] However, the same interviewee noted that the different changes are not easy to foresee and that there is a limited readiness to adapt to changes one is uncertain about. Any change in forest growth and extent must thus be seen in the context of public awareness as well as factors such as planting and soil conditions.

All in all, then, interviewees described a situation where it is not climate changes but the day-to-day and near-future economic situation that concerns them most. Most institutions do not have a climate policy or guidelines on the climate change question (Forestry and Reindeer Herding Administration, Sweden; Norrbottens Läns Skogsägare, Sweden; Metsähallitus, planning). Interviewees observed, however, that even climatic changes that extend trends already visible today could have notable effects on forestry. These effects include reduced site access and harvesting opportunities. Positive consequences would include increased tree growth and faster regeneration.

Table 3.2 *Main sensitivities and adaptations to climatic changes emphasized by stakeholders*

Projected climatic change	Local sensitivities and adaptations
Changes across the freeze–thaw line (onset of spring varying or earlier; milder winters with possible thaws; later autumn)	Impacts on load-bearing capacity of ground and roads. Adjustment costs; costs related to road building, maintenance, transport and access; impacts on production schedule and delivery times; increased needs for gravelling and sanding. Heavy wet snow may damage branches
Increased forest growth and longer growing season	Better seed crop and generally positive impacts of forest growth, although there will be changes in customer structure and price level in response to changes in wood quality; need for more frequent thinning
Shifts in tree species and spread	Changes in logging; potential limitation in spread of species due to soil quality

3.5. Conclusion: Vulnerability and adaptive capacity in forestry

Adaptive capacity to change in general – whether economic, political or environmental – is related to the numerous factors mentioned by interviewees as constraining their general situation and that have been described here under the headings of socioeconomic and political change. Interviewees' adaptive capacity and vulnerability to climate change cannot be assessed outside this context. The overall adaptive capacity of the area, especially in light of the low priority given to – and thus the limited planning for – climate change by actors, largely lies in the overall socioeconomic and political situation.

In general, when asked about their capacities to adapt, interviewees mentioned the socioeconomic/political problems they have experienced and often noted the adaptations they have already undertaken. They further pointed out why, for instance, their own companies or organizations still exist today (Common Forest II, Finland; Common Forest I, Finland). In many cases, it is the market system that provides both flexibility and constraints, for instance in demanding, allowing for and setting the economic limits on adaptation. On balance, the adaptations relate to company adaptation; possibilities of jobs in forestry in the future; adaptation to new owners, diversification, fuel and wood prices, and the overall market situation; environmental protection; and, finally, the context in which the sector operates, such as regional planning and municipal services and adaptation at large. Specific adaptations to climate change beyond those mentioned above form a small part of the overall response to change.

3.5.1. Adaptations at the company level and in employment

The market situation and companies' need to adapt to its demands were frequently discussed by interviewees. Market adaptations – including possible relocation, investment in technology and increased efficiency – were seen as part of continuous adaptations to international business cycles and technological development. The forest industry also needs to adapt to the situation in other counties; for instance, the fact that Stora Enso in Finland has sold its forest holdings has given rise to both concerns and new business ideas in Sweden (SCA, Sweden). This continuous adjustment in the industry, however, also results in some concern that large companies or units may leave the area (Metsähallitus, planning). Here, investments can be seen as adaptations to the market situation, since the industry will not be able to stay competitive without investment in new technology. Forestry also adjusts logging operations by calculating sustainable levels of logging that will assure future access to timber (SCA, Sweden; Common Forest II, Finland; Common Forest I, Finland). Here, one can speak about continuous adaptation to changes in the socioeconomic, political and environmental context.

At the level of company adaptation, on a smaller scale – such as that of sawmills – what was emphasized was the economic climate in terms of availability of resources and support. An export orientation aimed at accessing a larger market with more buyers was seen as a benefit. Other benefits mentioned were having a market rather than merely a production orientation to assure high added value by producing specialized products for high-paying customers and continuously renewing this network of customers (Sawmill Owner, Sweden). Similarly, access to private financing, for example through the parent company, was considered useful for reinvestment or for survival throughout difficult periods, inasmuch as financing for reinvestment, for instance, is difficult to obtain from banks or state loan institutions (Sawmill Owner, Sweden). Diversification by logging both spruce and pine was also seen as a way to maximize the industry's access to timber should one of these species become scarce on the market (Sawmill Owner, Sweden). The large prefabricated house company Älvsbyhus noted that its adaptations over time to the market situation have been the development of a good niche market and specialization, a limited organization with few managers, and focused development by, for example, selling off less profitable operations (Älvsbyhus, Sweden). Increased efficiency and rationalization have thus been the main means for the forest industry to remain profitable, but this has come at the expense of branches or actors that generate less profit and who have gone out of business because they were not able to use the competitive possibilities noted above.

Those working in the forest industry thus noted the need for preserving the industry as a viable employer in the future. Adaptations to this end would include thinning young forests and producing fast-growing energy wood to fulfil both current and future needs, using local timber for local buildings (to increase the use of local wood in Finnish Lapland), ensuring local access to timber (Sweden) and generally increasing refinement in the regions (cf. Lapland Regional Council, Finland). Forest smallholders also noted that the break-up of private forestry into small units, often with urban owners, presents a problem, as these owners are not as interested in forest management and may eventually decrease the value of their holdings (Forest Owners Union in Northern Finland). Small-scale actors in forestry, such as common forests, also suggested that they need to diversify as the focus on forestry decreases, for example by renting out land for cottages and tourism and increasing the area they own (thus benefiting from economies of scale) (Common Forest II, Finland; Common Forest I, Finland). In other words, some of the smaller actors emphasized the need for larger-scale organization. For instance, the Forest Owners Union in Northern Finland stated that it could help small forest owners cooperate to gain higher prices (Forest Owners Union in Northern Finland). On the whole, however, actors did not discuss local organization very much, as illustrated in the section on governance networks.

In the case of the forestry sector, it should also be noted that the overarching conditions are determined both by the market (including buyer and production structure) and fundamentally by the access to wood resources. Some actors noted that if access to wood became too unpredictable to maintain production at current levels, for whatever reason – even including the effects of weather on logging and transport conditions – a relatively small company without opportunities to diversify might have no choice but to close (Sawmill Owner, Sweden): 'Small sawmills have gone bankrupt, but that's the way it has always been here; one has died and a new one has appeared' (Lapland Regional Council, Finland).[115] A small Swedish sawmill noted that if local timber were to become more expensive, 'then we [would] close down, there [would] be nothing more [to do]' (Sawmill Owner, Sweden).[116] The larger and more profitable actors, on the other hand, have a larger adaptive scope; they could, for example, relocate if necessary because of increased timber costs or cheaper labour costs elsewhere.

3.5.2. Adaptation in the context of administration and policy

Actors in the forestry industry in the case study areas for the most part trained their description of adaptation on individual economic adaptations, as above. Some suggested that the municipalities could provide additional support for forestry companies and thereby employment (Local Forestry Association, Finland). However, the ways in which administration can support forestry are rather limited (Lapland Regional Council, Finland; Local Government II, Sweden). The principal forms it can take relate to planning, the maintenance of infrastructure, and, to some extent, the opportunities within the local municipality for services and diversification. For its part, the municipality can adapt to challenges impacting employment primarily by, for instance, developing education (possibly in cooperation with industry) or trying to attract new residents, industry and investment, the last possibly from international investors interested in the natural conditions of the area (Local Government I, Sweden; Local Government II, Sweden). The small communities studied here do not enjoy the benefits of scale that larger areas do: the situation of the inland communities can to some extent be seen as a problem of scale in that they are too small to attract a viable workforce. Lobbying politicians and forest companies to improve employment, for instance by inviting them to the municipality, may have limited impact: state forest organizations such as Sveaskog and Metsähallitus and private corporations may not acknowledge responsibility for employment even if they acknowledge that their operations have an impact on it (Local Government II, Sweden).

Another major problem in some of the inland areas with small communities is that expanding municipal businesses today, such as tourism, are seasonal and do not provide year-round employment and a stable municipal tax income; these, too, would be impacted by a changing climate and changing snow levels (Local Government II, Sweden). With the requirements for receiving unemployment benefits becoming more stringent, it is becoming more difficult for inhabitants to remain in their home communities with only seasonal work (Local Government II, Sweden). On the other hand, one local official noted that it has been the major external businesses that have dared to invest in the area, as they require the specific local environmental conditions and are not severely impacted by the long-term trend of an aging population and regional outmigration. Increasing adaptive capacity in the face of decreasing employment would require additional diversification in employment possibilities. Opportunities for this are limited but may materialize in increased wood refinement and state support for biofuel projects, for instance (Local Government II, Sweden). Where adaptation and differentiation are concerned, the difficulties lie in the many interests involved – for example efforts to develop tourism may disturb reindeer herding (cf. Local Government II, Sweden).

State decisions also make up the framework for many of the actions that impact the viability of communities: these measures include support through state loan agencies, the maintenance of public funding (for example military regiments and other facilities that support the community as a whole), and decisions on ownership of forest (such as Sveaskog policies) and environmental protection. Environmental protection is generally decided upon by the state and may impact forestry by limiting the available resources for logging. How actors would adapt to this is not clear, however: they mainly noted their own situations and adaptive capacities, and mentioned the things they would like changed as problems; few discussed specific changes, as if they considered these to be beyond their control.

Actors also suggested that the state should support forestry more through fuel subsidies, for instance, as one of the special features of northern areas is the need for transportation (Transport Entrepreneur, Finland; Common Forest II, Finland; Common Forest I, Finland; Metsähallitus, planning). Local actors are, however, constrained in their adaptation by the limited networks of communication and interaction and by much of the adaptive capacity being vested in external actors, including larger forest companies.

3.5.3. Particular adaptations to climate: Climate change as an addition to an increasing resource problematique

Particular adaptations to climate change include adaptations through road building, road gravelling, finding new customers for changing wood qualities, and possibly choosing seeds and planting in anticipation of change. Given today's business climate, with many closures during the 1990s, one may, however, expect that companies will look to the present more than to an uncertain future (especially those who do not own forest). Companies may see it as uncertain whether they will remain in the region in five years' time (or at their current size and performance level), which may limit any incentive to plan for forthcoming and uncertain changes. In sum, costs and a limited awareness of climatic changes impose constraints on these adaptations.

This conclusion shows that climatic changes must be seen in the context of the present economic and political situation and ongoing adaptations and priorities.

Despite decreased employment in forestry and a decreasing connection between the logging companies and the communities in which they operate, the timber harvest has in general increased. Yet, this increase has not been sufficient to cover the growing demand for wood. The result has been, among other things, heightened pressure on resources, increased competition among actors and between the forestry sector and other actors such as conservation, and a greater volume of imports from different areas. Interviewees found that the higher degree of regulation and additional demands on resources from conservation, for instance, could also reduce the area that forestry previously planned to harvest, constraining the resource basis for intensive forestry production. The decrease in available land for forestry was thus also seen as one cause contributing to resource scarcity; if the present conservation strategy had existed earlier, industry would perhaps have been able to adjust its activities and limit previous logging in order to maintain a constant level of extraction. This is a case in point illustrating the interconnected nature of adaptation over time that may act to limit or increase the scope for adaptation in the future.

When interviewees discussed the particular opportunities to adapt to climate, these tended to involve investment in better infrastructure (road quality). The difficult times are the transition times in autumn and, especially, spring. A shorter winter might be problematic because it limits access to logging sites and shortens the harvesting period. These problems will be manageable, however, if there is corresponding funding to adapt to the changes, for instance to provide resources for gravelling or improving roads. Pests and insects, which are today not prominent in the area, further constitute possible effects for which adaptations are not yet certain.

3.6. The nature of adaptation

Adaptation to date has limited potential adaptations in several ways and thereby increased vulnerability. This shows that vulnerability also needs to be considered in a historical perspective as the result of a resource regime or network of governance that has made some actions preferable to others. Earlier extraction of timber has limited the present adaptive capacity of actors; for instance, the state policies on logging, environmental protection and timber sales during different periods have impacted local access to resources. This illustrates some of the limitations in adaptation within an industrial sector whose logic is largely dictated by economics: sectoral interests may maximize harvest levels if this is possible within the regulatory and legislative framework even if this would constrain its future adaptations should the availability of resources decrease. The impact today of international environmental norms on the environment manifests itself in issues such as certification and environmental protection in nature reserves, also fundamentally affecting local communities. In this light, the current situation is best considered a result of historical adaptations to socioeconomic and resource conflicts, all of which create decision paths that often constrain present adaptation options.

The future of forestry in the areas studied here may be seen as determined by many factors, primarily economic but also political. The future of forestry will largely be determined by investment in the industry, which is to a large degree dependent on business cycles and access to wood resources. The vulnerability of local forestry actors is created and intensified within a multi-level framework, where the input and feedback mechanisms of the local actors to other levels are very limited. Local vulnerability is determined to a large degree by market constraints and national regulation. Restrictions exist in international market regulation (such as certification) and in institutional regulation, such as that imposed by the EU. These multi-level impacts govern and constrain the adaptation that can be undertaken at the purely local level. Figure 3.1, for instance, illustrates the range of actors that interviewees viewed as important across scale levels. There also exist few linkages from the local level to the actors whom interviewees regarded as the major influences on local management. Climate change is thus marginalized to a large extent in this system by limited awareness, problems felt to be more urgent than climate change and the possible costs of adaptation. Adaptations to climate change are also mainly undertaken or foregrounded by stakeholders to the extent that climate change impacts have an economic impact.

4

Perceptions of Change, Vulnerability and Adaptive Capacity among Reindeer Herding Stakeholders in Northern Norway, Sweden and Finland

4.1. Introduction: The organization of reindeer herding in Norway, Sweden and Finland

Reindeer herding is a small sector that must compete with the financially much stronger forest industry and other sectors for forest resources in what is a multiple-use context. The main income in herding comes from meat production, and here the sector faces competition from meat production in other countries, including farmed meat. Reindeer herding accounts for only a small minority of the workforce in the areas under study, meaning that the sector cannot argue for its importance for regional employment in the same way as forestry can. Nevertheless, as an occupation with long traditional standing and one often closely linked to the Saami (indigenous) identity, herding is often accorded special significance – even if only a minority of the Saami are today dependent on the livelihood. In Norway and Sweden, herding is mainly a Saami right, but in Finland all citizens have the right to practise it (Klokov and Jernsletten, 2002). However, in both Sweden and Norway there are concession areas where reindeer herding can be practised by non-Saami. In Sweden, for instance, non-Saami may also own a limited number of reindeer – known in Swedish as '*skötesrenar*' – but not herd them (Jernsletten and Beach, 2006). Reindeer ownership in such cases is subject to the approval of the relevant Saami village, which administers the reindeer for a fee, with the meat becoming the property of the owners.

This chapter first describes the reindeer herding situation in the case study areas in northern Norway, Sweden and Finland with a focus on trends common to the areas as well as notable differences between them. Some background information on the situation of reindeer herding in the countries, including organization of the reindeer herding sector and the general practice of reindeer herding, is provided in the following section. Section 4.2. provides an account of broad changes over time, especially with regard to technological changes and market adaptations. Here, increased supplementary feeding of reindeer with, for instance, hay or pellets, instead of only relying on natural grazing, has been used to attempt to compensate for encroachments from other sectors and the decreases in the labour available to reindeer herding owing to decreasing recruitment. Section 4.3. offers a description of the governance network for reindeer herding, which is centred on national regulation. International-level norms such as forest certification, which includes requirements on social criteria in forest production, and indigenous rights at large also have a large impact on reindeer herding. Section 4.4. describes ongoing and anticipated impacts of and adaptations to climatic changes in the area, which are more pronounced in reindeer herding than, for instance, in forestry; even relatively slight changes in weather have a considerable impact on reindeer grazing conditions, for example. The final section summarizes the vulnerability, adaptive capacity and particular adaptations discussed by interviewees in reindeer herding; these show that herding largely sees itself as having very limited possibilities for adaptation.

4.1.1. Reindeer herding organizations and practices

The reindeer pastures and the number of reindeer in Norway, Sweden and Finland are quite comparable (Klokov and Jernsletten, 2002). In Norway (in 2001), there were 165,000 reindeer grazing on 140,000 square kilometres, or about 40 per cent of the total area of the country; in Sweden, the figures (for 1998) were 227,000 reindeer grazing on 160,000 square kilometres, or about 34 per cent of the country; and the Finnish figures (for 2000) were 186,000 reindeer grazing on 114,000 square kilometres, or 33 per cent of the country. In 2000 Norway had about 2800 reindeer owners, in 1998 Sweden some 4700 and in 2001 Finland some 5700 (Klokov and Jernsletten, 2002). The number of reindeer for each country is limited by national regulations, mainly in order to control grazing pressure. Reindeer herding is practised throughout the northern regions of the countries, but is most extensive in the northernmost counties, which are the focus in this study. The diversity among reindeer herders in terms of herd size is substantial. For instance, Bostedt et al note that most Swedish reindeer owners own fewer than 50 animals. Moreover, not all owners live in the reindeer herding areas. They may have their reindeer herded by others and are thus not active themselves (Bostedt et al, 2001). On the other hand, the reindeer owners with large herds have an average of some 600 animals and comparatively large-scale reindeer meat production has developed on this basis in recent years. In Norway in particular, reindeer herding is comparatively large scale, with larger herds ensuring herders a relatively good financial situation. Herding is predominantly practised by men, and is in practice organized as small- to medium-scale businesses run by individual herders (Klokov and Jernsletten, 2002).

Administratively, the three different herding management systems in Norway, Sweden and Finland are based on similar principles. All three countries have established Saami Parliaments, which are decision-making consultative bodies of the state, but these only deal to a limited extent with the practical everyday regulation of reindeer herding and were rarely discussed by interviewees.[1] There exist a number of Saami parties as well as organizations, for instance those for reindeer herding Saami, although the only one of these discussed by interviewees was the Saami Reindeer Herders' Association in Norway (NRL). In all three countries, the general responsibility for national policy regarding reindeer husbandry rests with the ministry of agriculture. The ministry delegates the executive authority for implementing policy to different government services, such as the county administrative board (*länsstyrelsen*) in Sweden or the Reindeer Trade Authority (Reindriftsforvaltningen) in Norway. The Norwegian management system applies the concept of 'husbandry unit', a licence given to one person in order to legally own reindeer and to be part of a reindeer district. Each owner is an independent entrepreneur and a single family may comprise several individual enterprises. The Swedish system is similar to the Norwegian, but the licence is contingent on membership in a 'Saami village', which despite the name is not necessarily a village but refers to the administrative unit for the reindeer herding district. In Finland, a reindeer herding licence is connected to membership in a district, which in turn is a member of the Association of Reindeer Herding Co-operatives (also known as the Reindeer Herders' Association), the umbrella organization for the livelihood (Klokov and Jernsletten, 2002). One difference between the three countries is that while in Sweden and Finland a person is acknowledged as a reindeer herder through membership as above and owning reindeer, an individual reindeer herder in Norway is not eligible for support without also being designated a 'husbandry unit'. The Norwegian system thus includes an additional requirement in order to gain support as a reindeer herder. Traditionally, reindeer herding in all three countries was organized into units made up of several reindeer herders, who moved and managed their animals together. These units, called *siida*, are still used today by reindeer herders to indicate co-operative units within a reindeer herding district.

Reindeer herders today generally still graze their animals on natural pastures, maintaining summer and winter residences in order to keep track of the herd as it moves with the seasons. The movement follows the natural migration cycle of reindeer between summer and winter areas, the former season regularly spent inland and the latter in more coastal areas. The lands on which reindeer can graze are regulated in terms of the borders between different Saami herding districts and may sometimes also span national borders to reflect traditional grazing areas. In Sweden and Norway, the dates when reindeer can be on different lands are regulated by law, which breaks pastures down into summer and winter grazing grounds. In the Finnish system, there is no corresponding distinction: reindeer may graze within the herding district as natural conditions dictate.

In general terms, the reindeer herding year can be described as a succession of grazing, moving, gathering and slaughter periods. The exact timing of these differs according to area but can be broadly characterized as follows for the case study areas.[2] In the beginning of the calendar year reindeer are on their winter grazing grounds, subsisting mainly on ground lichen or supplementary feed such as hay or pellets where such feeding is practised by herders to extend or, in some situations, to replace natural grazing. When the snow starts to melt – in April or May (or earlier) – herders transport reindeer by lorry or drive them in herds towards the summer grazing lands using snowmobiles or, less often, helicopters. At the summer pastures, calving takes place in early summer (May and June in some areas) and the new calves are ear-marked in the latter half of June (Finland) or the beginning of July (Sweden). Ear-marking to distinguish ownership of each individual calf regularly takes place in summer before the summer heat and deerflies make reindeer scatter in search of forest shade or mountain areas with patches of snow. At this time, reindeer herders are able to take a vacation or work in supplementary occupations. In early autumn – September or so – many herders hunt. Following the hunting season, before the onset of winter, the older male reindeer are gathered for slaughter. There may also be some additional calf marking at this time and later in winter; the timing of this calf marking varies across the areas. In September and October, reindeer herders may also work in additional occupations, repair reindeer fences and so on, and take care of administrative tasks and meetings, for instance consultative meetings with forestry regarding logging, as reindeer herding is generally practised on land owned by state or private interests where reindeer herding rights are based on traditional use. After that, the herds are monitored and gathered together, with the reindeer from different herds separated for the subsequent movement of herds to the winter grazing grounds, which often occurs sometime in November: 'At that time we have to gather and separate and slaughter, during that darkest, worst time. That's when we have the most work … in October, November, December' (Reindeer Herder I, Sweden).[3]

Significant regional differences can be seen in reindeer herding within as well as between the countries. Some of these reflect different national reindeer herding systems; others reflect differences between regions, regional choices, and regional (natural) constraints and available pasture; and still others depend upon whether reindeer herding takes place mainly in forest or mountain/open areas. Regional differences can also be a result of the scale of forestry in the area, which is extensive in the case study areas in Sweden and Finland and has a decisive impact on the grazing resources available for reindeer herding, but is essentially insignificant in the case study area in Norway. Moreover, in different areas, spring arrives at different times, affecting, among other things, the timing of spring migration (Reindeer Herder III, Sweden). Inland areas may have more mosquitoes than coastal areas (such as in Norway) do, which also affects grazing conditions (Reindeer Herder II, Norway).

As mentioned above, there are also differences in herd size not only between individual herders but also between regions and countries. In Sweden, for instance, reindeer herders in forest areas (*skogssamebyar*) note that an individual herder needs to have 'at least 500 reindeer [if he or she is to make a] living from reindeer herding only' (Reindeer Herder I, Sweden).[4] In Finland, reindeer herders would need some 300 reindeer to make ends meet, but often own fewer animals than that, and may even be considered large owners if they have 400 reindeer or over. In Norway, it is easier for established herders to make a living from herding alone as reindeer herding is regularly larger in scale (Reindeer Slaughter House, Norway). This difference may be due in part to the administrative organization, in part to the more dominant position of reindeer herding in the country and also to the natural conditions. In Finland and Sweden, reindeer herding takes place largely in forested regions, which requires more work. The open spaces and comparatively low population density in the case study area in Norway make it easier to herd and gather reindeer (The Reindeer Trade Authority, Norway).

4.2. Socioeconomic changes in reindeer herding

Reindeer herding practices have changed considerably during the last generation (20 to 30 years). To some extent this is a reflection of the aging population structure in the areas and to some extent an indication of increasing market pressures and competition from other sectors and of technological advances. A person who had worked as a reindeer herder since 1970 observed that the environment for reindeer herding has changed in the form of diminishing pastures and increases in other activities on pastures, increased mechanization, stabilization of the meat production through, among other things, supplementary feeding of reindeer in addition to natural grazing, and increased costs for the livelihood (Reindeer Research Centre, Finland). He noted that 'profitability is perhaps the [main problem] at present ... The structure of costs and income is not good. Costs have risen so high in comparison to income and there are many reasons why the reindeer and the herders have changed – the whole reindeer herding system and the environment' (Reindeer Research Centre, Finland).[5] Similarly, in the opinion of a Swedish reindeer herder, there are two substantial problems:

> *The poor profitability – that is absolutely problem number one. Problem number two is the encroachments from every direction ... It is not only forestry that is reducing the size of the grazing areas but all other encroachments that add up ... It is the airport, the car testing, tourism.* (Reindeer Herder I, Sweden).[6]

Together these different pressures may, as one actor assessed, cause reindeer herding to lose up to 50 per cent of its pastures in some areas as compared to the land it had a generation ago (Reindeer Herder III, Sweden). In particular, old forests have been logged and spruce copses removed: '"rubbish forest" as the forest workers themselves put it. As far as we were concerned, it was not rubbish forest that disappeared but forest with arboreal lichen' (Reindeer Herder I, Sweden),[7] which reindeer herding uses as a traditional type of emergency fodder. In the Norwegian case study area, impacts on reindeer herding often come from the building of cottages and other infrastructure development rather than forestry, which is not practised in this northern and largely coastal area.

All of these issues, together with increased reindeer herds, impact pasture conditions and the long-term viability of herding, and will be discussed in turn below. Reindeer herding today exists within a complex where limitations on grazing resources, requirements of increased economic efficiency and technological possibilities prompt economic adaptation towards larger-scale herding, sometimes with an emphasis on feeding reindeer. The issue of economic profitability and a functioning meat market was emphasized in all three areas.[8] The problems in Norway are generally less extensive than those in Sweden and Finland, however, as reindeer herding has a stronger position in the economy, is practised on a larger scale and enjoys a somewhat stronger market for reindeer meat. Moreover, herding operates in a different environment, at least on the coast, and does not have to compete with forestry, which is not practised commercially in the case study area.

4.2.1. Increased mechanization, costs and supplementary feeding

To some degree, the current problem of limited profitability has resulted from the increased mechanization since about the 1960s, which has raised the costs of reindeer herding (cf. Reindeer Research Centre, Finland). 'Greater demands have been placed on the profession ... to make it [function] economically ... there is more work and higher expenses' (Reindeer Herder III, Sweden).[9]

> *It has become mechanized and more technical all the time; the costs have changed and the feeding ... all this has increased costs. Medication for the reindeer and such. Of course this has had positive effects as well; it has brought [good things] not just taken things away.* (Reindeer Research Centre, Finland)[10]

The demands of economic profitability and the costs have thus both grown: one actor said that when he started some 25 years ago, herders rented empty houses instead of maintaining permanent winter residences. Today, herders regularly own and maintain permanent winter as well as summer homes (Reindeer Herder III, Sweden).

Technological developments have also been similar in the three countries. Some 25 years ago, herders skied when moving the herd to summer grazing lands, using perhaps only one or two snowmobiles. However, they lost much of the herd, as animals went astray in the forests, for instance (Reindeer Herder III, Sweden). Today, lorries are used for transport and helicopters and snowmobiles for herding, which increases costs but also results in less of the herd being lost and thus opportunities to increase meat production (Reindeer Herder III, Sweden). The herders also use 'two-wheeled motorcycles or four-wheelers ... there is probably no one who walks those distances now as we [did] before' (Reindeer Herder III, Sweden).[11] In the Norwegian case study area, reindeer herders use motocross motorbikes, the preferred transport in an open rather than forested landscape (Reindeer Herder II, Norway).

The use of different technologies is also considered necessary for today's economically efficient reindeer herding, where the maximum number of reindeer is nationally regulated: 'Reindeer herding can be carried on economically despite everything because you can use snowmobiles and even motorcycles' (Reindeer Herder III, Sweden).[12] Turnover has increased many times over, but may to some extent have been matched by increases in costs: 'The costs increase all the time, above all where fuel is concerned as we have noticed; then there is everything else that goes into [herding] – snowmobiles, flights, helicopter costs' (Reindeer Herder I, Sweden; cf. Reindeer Herder III, Sweden).[13]

In Sweden and Finland, interviewees mentioned the recent increase in the use of helicopters (to increase the efficiency of herding in forested areas) in particular. One Finnish reindeer herder, for instance, noted that a large number of reindeer can be gathered by helicopter as compared to other methods; however, he felt 'the reindeer will get used to the sound of the helicopter sooner or later' and not be driven as easily by helicopters as they are now (Reindeer Herder II, Finland).[14] A Swedish reindeer herder noted that his use of a helicopter was to some extent a result of the plantations of quick-growing contorta pine, which grow very close together and make it more difficult to gather reindeer (Reindeer Herder II, Sweden). Although the use of a helicopter is more costly than other means, it also makes it easier to see the reindeer in the dense plantations (Reindeer Herder II, Sweden). This Swedish reindeer herder thus uses the helicopter 'summer, autumn and winter, especially when there is little snow and I cannot go by snowmobile' (Reindeer Herder II, Sweden).[15]

Some herders noted, however, that the increasing level of mechanization, although largely considered beneficial, may result in a loss of reindeer herding knowledge. A Finnish reindeer herder asked:

> *What are we going to do when this technology becomes tried and true if we then get a new generation of reindeer herders who do not know how to take care of reindeer? They won't know how to find them, how to gather them or transport them, or how to keep them in a group during transportation. This will lead to massive difficulties.* (Reindeer Herder II, Finland)[16]

Indeed, one herder suggested that reindeer herding 'could "hit the wall" if it doesn't bring itself together and start to cut down on technology and those aids' and in this way maintain traditional herding knowledge (Reindeer Herder I, Norway).[17] Another actor suggested that technological developments, if they continue bringing higher costs, might 'mean a concentration of reindeer ownership' in the hands of relatively few individuals (Employment and Economic Development Centre, Finland).[18] This trend poses a multifaceted problem. As the costs rise without income rising proportionately, fewer individuals are able to make a living through reindeer herding. This means that herds have to be managed with fewer personnel, resulting in an even greater dependency on technology. The economic situation and the limited room for new recruitment combine to create a situation in which few new reindeer herders start out (Reindeer Herder III, Sweden; cf. Employment and Economic Development Centre, Finland). This situation may, in the long run, limit herding's viability as a livelihood, as fewer and fewer young reindeer herders go into the business:

> *I view the future of reindeer herding rather darkly ... A large problem is the recruitment, the space is limited ... as the development is now, we are progressing towards larger and larger companies, more and more reindeer, and then we have the limiting factor of grazing area. You cannot have however many reindeer you want. And for each company ... if you are going to find a level where you can make a living you cannot have an unlimited number of companies. And at the same time you cannot have too few [companies] if you are going to practise effective reindeer husbandry. So it is really difficult to take in new people in that situation who want to go in[to the profession] and commit themselves because it must be horribly difficult to start up ... I almost think you need to inherit a reindeer herd.* (Reindeer Herder I, Sweden; cf. Reindeer Herder III, Sweden)[19]

The problem with recruitment is among the issues most heavily emphasized in Sweden and Finland. It is regarded as a result of many factors, foremost among these the economic situation: the limited financial opportunities due to low meat prices, as well as the limited support for start-up businesses and pensioners leaving reindeer herding, are seen as causing family problems and making reindeer herding a less desirable occupation. Because of problems with profitability, many herders also require a secondary income, especially in Sweden and Finland. As herders noted: 'A few companies make it on reindeer herding only but many require additional opportunities for work' (Reindeer Herder III, Sweden; Reindeer Herder I, Sweden).[20] The jobs that reindeer herders take on may be, for example, thinning or other silvicultural measures in forestry (Reindeer Herder III, Sweden; Reindeer Herder II, Sweden; Reindeer Herder I, Sweden), work in tourism, driving lorries or taxis (Reindeer Herder I, Sweden), or machine repair (Reindeer Herders'

Association, Finland). A large part of the income of reindeer herding families comes from the woman in the family working outside the livelihood and thus enabling the family to cover some of the losses that may be incurred there.

In Norway, there seems to have been more of a generational shift that has prompted new herders to come into the business (Reindeer Slaughter House, Norway). One problem affecting the livelihood there is what people perceive as a policy at the municipal and higher levels of reducing the number of reindeer husbandry units (*driftsenheter*), which limits the positions available for those who want to continue in reindeer herding and steers the sector towards larger but fewer reindeer owners, 'in a way a centralization of rights' (Reindeer Herder I, Norway).[21] 'That means that there are rights that are taken away from others who would like to practise reindeer herding' (Reindeer Herder I, Norway).[22] This situation has caused concerns that reindeer herding skills will be lost.

At the same time as the percentage of the population involved in reindeer herding is diminishing, reindeer herding needs to interact with a large number of parties, including local and regional administration and other forest users. Reindeer herders today need to coordinate their work with all other areas of land use, as municipal planning and change are taking place relatively quickly (Reindeer Herder III, Sweden):

> *Quite a change has taken place here in reindeer herding as a livelihood in 20 years in that the amount of work of the area administration here has increased. Anyone can see what I mean and you yourself can imagine how you have to be involved in almost anything and everything that has to do with the livelihood.* (Employment and Economic Development Centre, Finland)[23]

Thus reindeer herders need to be aware of 'environmental issues, mountain management, hunting, fishing, snowmobile questions, forest management questions; there are regulations – hunting regulations and such' (Reindeer Herder III, Sweden).[24] This means that a smaller population needs to interact on a larger number of issues with an increasing number of actors. 'There has been, I would say, an enormous change from when we walked and skied in the winter – in such a short time as 25 years' (Reindeer Herder III, Sweden).[25]

4.2.2. Supplementary feeding

Given a situation with profitability concerns, limited human resources to undertake the work and extensive technological changes, one adaptation to improve efficiency has been to provide fodder to reindeer during particular periods of the year rather than rely exclusively on natural grazing. In Finland and Sweden, where reindeer herding is the most economically pressed, supplementary feeding using hay or pellets has increased significantly in the last decade. Supplementary feeding makes it possible to maintain larger herds than could otherwise be sustained on limited pastures (often because grazing has been limited by logging or other disturbances) and helps control the variations in reindeer numbers and stabilize herders' incomes. The increase in supplementary feeding is thus to some extent a result of the logging of old forests and other restrictions on natural grazing, and to some degree a result of increasing reindeer numbers, demands of the market and changes in practices. Today supplementary feeding is undertaken in periods when grazing is unavailable or reduced, and in some areas – especially in Finland – herding actually takes the form of reindeer farming, in which the animals are fed in enclosed areas rather than being allowed to follow their natural migration and grazing patterns.

Accordingly, in Finland, 'there have been enormous changes in the last ten years and some of the herders try to hang on to the profession by grazing reindeer on natural pastures or feeding them in the wilds; but some have clearly begun to keep them in pens' (Reindeer Research Centre, Finland).[26] Reindeer herders in the case study area in Finland often own land, and may have access to hay harvested on it. Supplementary feeding has, among other things, increased reindeer meat production. A Finnish reindeer herder noted that in the 1970s the average calf weighed only 18 kilogram, as compared with a current weight of 23–24 kilograms: 'Over 30 kilos is not unusual for a calf' (Reindeer Herder I, Finland);[27] 'that is from supplementary feeding' (Meat Buyer, Finland).[28]

In Sweden, supplementary feeding has also increased, but at least in the case study area it was not discussed as a break with traditional reindeer herding or as a form of farming as it was in some cases in Finland. Rather, supplementary feeding was used in grazing emergencies to an increased extent, mainly because of the decrease in the natural emergency fodder of arboreal lichen now that old forests have been logged. Supplementary feeding takes place at times when the ground lichen stands freeze over or are otherwise inaccessible. The level of and need for feeding is 'highly variable' (Reindeer Herder III, Sweden):[29] 'It depends on the composition of your grazing grounds ... those with limited winter grazing grounds and who have had a very high level of logging on the areas [such as us] – they perhaps have to start [supplementary feeding] earlier' (Reindeer Herder III, Sweden).[30] It is 'especially in late winter and spring that we have to feed them and also during bad grazing conditions in winter since there is no old forest where they can graze on the lichen' (Reindeer Herder III, Sweden).[31] This supplementary feeding results in increased expenses for reindeer herding. A Swedish herder estimated that in his Saami village of fewer than ten herders some 60 tons of extra fodder are used per year, 'which we never used before when we could release the reindeer into the lichen forests' (Reindeer Herder III, Sweden).[32] For a Saami village with about 3500 reindeer and fewer than ten herders, one interviewee noted that:

> *The lack of arboreal lichen grazing in the spring period has resulted in our having to provide supplementary feeding during this period after*

> *we have migrated up [into the mountains]. And then you can say that this has a direct economic effect because it affects our capital; we have to provide, say, supplementary foddering to the tune of [SEK] 250,000 each spring ... right out of our own pockets.* (Reindeer Policy, Sweden)[33]

In Sweden, the cost of supplementary feeding is thus substantial: 'We buy [the fodder] ... it is pellets, that is the only saving grace. We do provide supplementary feeding also with ensilage ... that we buy from farmers in the coastal areas here' (Reindeer Herder I, Sweden).[34] While there is a possibility to gain state support amounting to 50 per cent of the cost of supplementary feeding in the case of grazing crises, in one Swedish herder's opinion, 'Even if you get this catastrophe relief you have to put up 50 per cent yourself so that is why you hardly make it' (Reindeer Herder I, Sweden).[35]

In Norway, by contrast, supplementary feeding is relatively uncommon, at least in the case study area. 'It is many years since we provided supplementary feeding. We did it during the bad years ... and then we only fed them out on the mountain. We did not have them in a pen or anything like that' (Reindeer Herder II, Norway).[36] One reason a Norwegian herder mentioned for why supplementary feeding had not been necessary was the good climatic/weather conditions for reindeer herding over the previous few years and the fact that herders have grazing rights in areas used during spring migration (Reindeer Herder II, Norway).[37] That some reindeer herders in the case study area have also started supplementary feeding was viewed both positively and negatively. 'Some want to give supplementary feed, and others don't ... we have split up into different *siida* and that is great. Then we can ... use rather large areas here and ... those who use supplementary feeding take smaller ones' (Reindeer Herder II, Norway).[38]

The cost of supplementary feeding in Norway is thus reasonable because of the limited feeding and the relatively low cost of fodder, leaving the Norwegian case study area with little impact from supplementary feeding. Rather than buying fodder, 'most people [in the area] actually harvest hay themselves. There are people who have some property because of the fishing rights ... I do not think that we'll ever come to the point where we provide supplementary feeding through pellets and concentrate fodder' (Reindeer Herder II, Norway).[39] Reindeer herders also note that supplementary feeding may result in decreased meat quality. Allowing reindeer to graze naturally could thus become a competitive advantage:

> *I think that there is a difference whether you feed them in a pen or out in the wilds. In a pen they become a bit like farmed fish. At the same time, we don't dare emphasize this too much. We don't dare go out actively and criticize the quality of meat because we know that if there is mild weather in December we, too, will have to give the animals supplementary fodder.* (Reindeer Herder II, Norway)[40]

4.2.3. The meat market, meat prices and profitability

As noted above, reindeer herders view profitability as a major concern. In some cases, reindeer herders may sell meat directly to friends and neighbours locally or through other networks to augment their income, but this is most often a limited option (Stakeholder meeting, reindeer herding, Finland).[41] Despite the adaptations mentioned above, reindeer herders are not able to control the factor that perhaps impacts profitability the most – the market price of reindeer meat. Interviewees estimated the decline in meat prices during the three years preceding the study at 13 per cent in the Norwegian case study area and some 25 to 30 per cent in Finland (with Sweden falling between the two) (Employment and Economic Development Centre, Finland; Reindeer Herder II, Norway; Reindeer Herder II, Sweden). As meat sales constitute the main source of income for reindeer herders, the drop in price had repercussions for the entire sector. The development can be attributed to multiple factors, including changes in the import–export situation between countries, lower demand for reindeer meat products by consumers and competition with other game meat, such as New Zealand deer. Herders noted that there also seems to be a market imbalance, since 'for the consumer [the reindeer meat] has not become cheaper in the shops' (Reindeer Herder I, Sweden).[42] However, interviewees noted that it is difficult to determine the exact factors behind the price decline and how they interact: 'There has not been a proper study so that somebody could compare things' (Reindeer Herders' Association, Finland).[43]

These concerns about falling meat prices were shared in Norway, Sweden and Finland, although Sweden and especially Finland have been the most seriously affected given their smaller-scale reindeer herding. A general underlying reason for the drop in meat prices noted by interviewees is the market situation. For instance, it has been difficult to find buyers for all the meat, which lowers prices (Meat Processing, Finland; cf. Stakeholder meeting, reindeer herding, Sweden). In a situation of limited demand and few buyers, prices fall as the meat supply increases, as has occurred in the last few years, when environmental conditions have been beneficial (Employment and Economic Development Centre, Finland). The situation is especially pronounced in Finland, where the reindeer herding units are relatively small; as one interviewee noted, it takes almost as much in the way of resources to maintain a 100-head as a 300-head herd, but the economic benefit of the latter is greater (Reindeer Herder I, Finland). In Norway, there is also a meat surplus because there are some 75,000 reindeer too many in Finnmark (more than a third more than the national limit). The reasons for this include 'very beneficial conditions. Mortality is low; there are few predators; there are natural conditions such as the mild, favourable winters ... these have been banner years' (Reindeer Slaughter House, Norway).[44] In this respect, reindeer herders may be seen as caught in something of a Catch 22: they have continuously adapted through efficient technologies and supplementary feeding to maintain a stable and high level of production, but when they are too successful and, in addition, are supported by beneficial environmental conditions, they may exceed the limits imposed on the number of reindeer, causing meat prices to fall (Saami Parliament, Norway). Environmentally beneficial conditions may thus contribute to the present crisis in what is a limited meat market (Reindeer Herders' Association, Finland).

In Sweden, the drop in meat prices was caused by certain regulatory mechanisms in addition to the market situation: 'Firstly, price support was halved by the end of the year. In addition, the price ... decreased catastrophically' (Reindeer Herder I, Sweden).[45] In Sweden, the increase in reindeer herd sizes has, however, not been as marked as in Finland and especially Norway (to some extent as the weather conditions in Sweden have not been considered as favourable). The Swedish market, however, is impacted by the Finnish and Norwegian markets, among others, where surpluses are depressing prices. At the same time, Swedish herders are not able to increase the number of animals slaughtered very much (even if they could sell the meat), as herd sizes are not as high as in Norway, for example. A Swedish reindeer herder noted that:

> *There are no possibilities to increase slaughter outtake to compensate for the drop in prices. You try to aim for the maximum; every entrepreneur slaughters as many animals as he can, what he judges is possible to take out of the herd. If I overdo it, then capital is decreased and that is not sustainable in the long run unless of course you try to come up with supplementary income.* (Reindeer Herder I, Sweden)[46]

On balance, the price of reindeer meat fell somewhat in the last few years before the study as a result of surplus production under beneficial weather conditions. The relatively limited reindeer meat market has not been able to handle the resulting production, with detrimental consequences for the livelihood. Reindeer meat is often exported to neighbouring countries (for instance when a surplus exists beyond what is purchased by regular buyers in the home region). When Norwegian slaughterers and producers stopped buying reindeer from Finland because of the overproduction in Norway and, to some extent, a rather high customs duty on imported meat, the reindeer herding areas in Finland lost an important outlet that had paid good prices. Finnish buyers were subsequently able to hold out for lower prices and still purchase meat, as Norwegian buyers withdrew from the Finnish market (Reindeer Herders' Association, Finland; Reindeer Herder II, Norway; Reindeer Slaughter House, Norway; Reindeer Herder I, Finland). There has also been import of reindeer meat from Russia (Stakeholder meeting, reindeer herding, Sweden). With few established buyers and established sales networks, competition becomes limited, outlets are few, price levels decline as the supply increases, and sellers have to compete to get the meat sold (Meat Processing, Finland). For instance, 'many reindeer owners associations are in a situation where there are no buyers for their reindeer' (Reindeer Herder I, Finland).[47] Actors also noted that the limited competition between buyers and small networks sets a common price level: 'There is communication between different buyers and they know how things are – there is no real competition between them' (Meat Processing, Finland; cf. Stakeholder meeting, reindeer herding, Sweden).[48] As a result, 'last year there were suspicions that a cartel had been formed' (Meat Processing, Finland).[49]

One final reason brought up for the drop in reindeer meat prices was competition with other meat, primarily New Zealand red deer. Its price on the supermarket shelf (two euros per kilogram) is less than half of what meat buyers pay reindeer herders for meat on the hook (at the time of the study some 2.5–4 euros/kg, as compared with a shelf price that may be as high as 50 euros/kg) (Meat Buyer, Finland; Meat Processing, Finland; Reindeer Herder II, Norway). 'It is hopeless to compete with [red deer] in price' (Reindeer Herder II, Norway).[50] Among the reasons for the price difference are that the production costs for New Zealand red deer are lower and the quantity of meat produced is higher: the animals grow large in a favourable climate (Meat Buyer, Finland; Meat Processing, Finland). The low price of red deer meat attracts consumers and even those who normally buy reindeer meat purchase it: 'It is so easy to see it when one goes to the frozen food counter where we have both red deer and reindeer meat – guess which package the consumer usually chooses? Many people take the red deer and think "it's only stew meat, anyway"' (Meat Processing, Finland).[51]

In addition to their apprehension concerning competition with New Zealand red deer meat, interviewees also perceived a risk that reindeer from Russia or venison from Estonia, Latvia, Lithuania and Poland might put increasing pressure on game prices (Meat Buyer, Finland; Reindeer Herder I, Finland). In the opinion of one interviewee, another problem is that most companies are only interested in stew meat and do not want to try new reindeer products, where the quality of reindeer meat is superior to that of red deer (Meat Processing, Finland).

4.2.3.1. Solutions to the meat market problem: Refining and marketing to increase consumption

Reindeer herding has largely been placed in a situation where herders' own individual capacity to adapt is limited. Given this situation, one key approach that interviewees saw for dealing with the problem of declining meat prices was to increase the consumption of reindeer meat:

> *In the 1990s the [reindeer meat] production was at a low ebb ... so people in the consumption centres actually forgot about serving reindeer on everyday and festive occasions. This has certainly been a problem for both the reindeer owners and the buyers, and now they have started to do something about getting people to eat reindeer again – and that is the best remedy in this situation ... the best guarantee of a decent price.* (Reindeer Herders' Association, Finland; cf. Reindeer Slaughter House, Norway)[52]

In Norway, one additional reason perceived for the decrease in reindeer meat prices was the decline in the advertising of reindeer meat, which interviewees associated with changes both in the administration responsible for marketing reindeer meat and in consumption patterns (The Reindeer Trade Authority, Norway):

> *You are not able to market today ... We must have TV and we must have large actors who market and who come in to make a profit. It is not like before that people come home to you to buy reindeer meat. They do not do that anymore. Now it has to be in the store and there have to be small packages and there have to be directions for use on them.* (Reindeer Herder II, Norway)[53]

Both advertising campaigns and modifications in processing were suggested by actors as means to alter consumer habits and increase demand (Saami Parliament, Norway). In Finland, for instance, new packaging and product packages for reindeer meat have been introduced in the supermarket aisles in cooperation with supermarket chains, along with recipes for reindeer meat. This has been achieved through cooperation between the reindeer herders' association and supermarkets without using intermediaries (Local Government, Finland; Meat Buyer, Finland). In Norway, some interviewees also envisioned a similar system (Saami Parliament, Norway). A focus on exports beyond neighbouring countries was also discussed by interviewees as part of a need to target the broader international market (Reindeer Herders' Association, Finland; The Reindeer Trade Authority, Norway). Norwegian interviewees noted that with the tax levied on Norwegian exports to the EU, the country should target markets beyond Europe, such as Japan and the US (Reindeer Slaughter House, Norway).[54]

However, some obstacles to the broader marketing of reindeer meat remain. These include the development of a working system for the transportation of live reindeer and reindeer meat (Meat Buyer, Finland). Freight to the south is expensive (Meat Buyer, Finland). The EU requirements concerning a certain standard for slaughterhouses, which so far have entailed added costs, may in this case support export (Reindeer Herders' Association, Finland; Meat Buyer, Finland). However, the seasonal nature and location of reindeer herding production pose obstacles to trade. While suppliers would like to see reindeer meat on the shelves year-round, reindeer are mainly slaughtered in autumn before they scatter and start losing weight in winter. Moreover, reindeer meat is a relatively limited product: supply may be too limited and prices too high for a larger export effort (Reindeer Herder I, Finland).

Finally, to improve the system over the longer term, it has been suggested that meat prices should be stabilized by cooperation between the countries, with buyers and sellers agreeing on who the buyer will be before the reindeer are slaughtered (Employment and Economic Development Centre, Finland).[55] On the other hand, some reindeer herders noted that there is a risk in developing market cooperation further. In Finland, it was noted that, as 75,000 reindeer need to be slaughtered in Norway to bring the number down to nationally determined limits, 'if this meat came to the Finnish markets, there would be a lot of problems' (Meat Buyer, Finland).[56] A Norwegian reindeer herder noted: 'We know that the prices in Finland and Sweden have only fallen, and fallen the entire time, and we have had this customs duty protection, but if the prices fall in Finland and Sweden even more, it may be profitable to import [meat] paying full customs duty; and they may be doing that already' (Reindeer Herder II, Norway).[57] Slaughterhouses corroborate this: it would have been slightly cheaper, even with the customs duty that has to be paid to Sweden and Finland as EU members, to buy reindeer in Finland (given the low Finnish meat price). The reason that this did not happen is the sharp criticism it would have engendered, which in turn might have lowered sales and worsened buyer–herder relations on the Norwegian side: 'Norwegian reindeer herding would shoot itself right in the foot' (Reindeer Slaughter House, Norway).[58] On that account, Norwegian domestic production was prioritized over import (Reindeer Slaughter House, Norway).

In sum, then, reindeer herding can be seen as heavily influenced by its economic environment. Interviewees observed that the requirements of the market economy affecting reindeer herding have increased during the time they have been working. This has heightened demands for economically efficient production and supplementary feeding, whose aim is to secure an income from reindeer herding and to compensate for disturbances from other sectors that restrict the use of pasture lands. The meat market and price levels were the issues most frequently discussed in the interviews, with the topics taken up including the meat surplus, the limited number of buyers and the need to increase consumer-oriented actions such as the marketing of reindeer meat to raise demand. Interviewees noted that the meat market is influenced by, among other things, competition with deer meat from New Zealand and by EU taxation borders and regulation. The political and economical sectors are thus to some degree interconnected. A Swedish reindeer herder expressed the common problem for reindeer herding across countries: 'Today the situation is that it is regulated … that you can have a certain number of reindeer, full stop, and the deal is to gain maximum yield from them' (Reindeer Herder III, Sweden).[59]

4.3. Political change in the sector: Legislation and support

This section describes the national regulation system for reindeer herding as regards support and herd size limitation, the external impact on reindeer herding from other sectors, and the impact of EU and international norms on the livelihood. The description will start from the national level of regulation, the one at which most of the regulatory mechanisms are located. National regulations also prompted extensive discussions in all areas studied of the systems and limits of support, reindeer number limitations, and whether the current meat price problem could be dealt with through increased support or state subsidies for additional slaughter to help herders comply with limitations on herd size. The discussions of national support and support systems became topical especially due to the decline in meat prices and the need to compensate for this. Herders noted, however, that they have little influence at this level.

Also taken up below are the sectors that interviewees saw as competing with reindeer herding. These sectors – especially forestry – were regarded as having a large external impact on reindeer herding and, despite some coordination, greatly reducing the space available for the livelihood. Some of the discussion also dealt with the EU level of decision-making, as it impacts slaughter regulations and has resulted in new slaughterhouses being built. The principal impact at the international level can, however, be attributed to International Labour Organization (ILO) Convention No 169 regarding the rights of indigenous peoples, which at the time of the study Norway had already ratified and Sweden and Finland were examining for potential ratification. The convention provides one means for reindeer herding to argue for its rights relative to other sectors. On balance, reindeer herding can be seen as comprising a relatively limited actor, one removed to a large extent from possibilities to provide feedback on national regulation. As a small economic actor, herding is also greatly affected by other sectors and coordination with them; moreover, it has limited power to influence such coordination. It is in this connection that international norms such as the ILO Convention have become especially important, as they can provide leverage vis-à-vis the state and other actors.

4.3.1. National systems for financial support and stabilizing the number of reindeer

In all three countries, there exists a state-determined highest number of reindeer, which is in principle based on the assessed carrying capacity of the reindeer herding areas. The national support (financial support or subsidy) systems can be said to serve dual purposes: they are designed to both support reindeer herding and stabilize the number of reindeer at determined levels, for instance by encouraging slaughter through slaughter support. There are thus complex linkages between regulations on the size of the reindeer population, support systems and herders' adaptations to maximize output from the herd (for instance through supplementary feeding), especially given the limited available territory for the herd.[60] The main support systems provide financial support either per live reindeer kept over the winter (Finland) or slaughter support (Sweden and Norway). There are also general support systems, depending on the country, for young reindeer herders just starting out, pension benefits, and emergency support if grazing areas freeze over or become inaccessible in any other way for longer periods. The main discussion concerning support systems in all three countries dealt with levels of support and whether one's own country should adopt the support policy practised by one of its neighbours. There was thus a pronounced focus on comparisons between countries of support and market systems, among other things, which might level the playing field where reindeer meat sales are concerned. Reindeer herders also compared the support for reindeer herding with that in other sectors: 'If you compare [herding] with agriculture – the average farm with the average reindeer herding company – the average farm gets four times the support' (Reindeer Herder III, Sweden).[61]

In Finland, some actors suggested that a slaughter support system could, for example, be possible as an addition to the current system (Employment and Economic Development Centre, Finland). Quota systems to better control the number of reindeer and keep it below the established limits have also been discussed (Employment and Economic Development Centre, Finland). However, actors also suggested that it may be 'difficult to determine whether the present system or a quota system is better' (Employment and Economic Development Centre, Finland).[62] Other interviewees suggested that additional support systems could be instituted, for example that reindeer herding should be compensated for disturbances from other sectors, or that the state should subsidize reindeer meat sales (Reindeer Herders' Association, Finland; Employment and Economic Development Centre, Finland). Some interviewees, however, argued that reindeer herding 'cannot be built on subsidies'; it needs to find ways to develop economically as a sector (Reindeer Herder II, Finland).[63]

Given the limit on the number of reindeer and the Finnish pattern of many small-scale reindeer herders operating over relatively small areas, reindeer herders continuously adapt to maximize reindeer herd productivity. One form this has taken is to change the herd structure by slaughtering most males in autumn and retaining productive females to produce next year's calves in the winter herd, the size of which is the basis for the highest permissible number. Another way to raise productivity has been supplementary feeding, which has increased considerably in some areas and which serves to help the entire winter herd survive the winter and to obtain more meat per reindeer. Yet this adaptation has resulted in concerns about growing reindeer herd sizes. Those who engage in supplementary feeding increase their productivity and thus improve their financial situation but this comes, at least to some extent, at the expense of others who herd reindeer traditionally and whose productivity is subject to natural constraints such as grazing limitations. Since in Finland there is a slaughter percentage required of all herders, those who have herded traditionally may need to slaughter their productive animals to meet their percentage, while those who 'farm' animals may only have to slaughter those that are not a loss in terms of herd structure and thus succeed in increasing rather than reducing their herd sizes. This situation may result in competition and struggle among reindeer herders within the district to maintain productivity and thereby maintain their place in the district. There is thus a risk that large-scale reindeer herding (even though reindeer herding in Finland is relatively small in scale compared to that in Norway) will expand at the expense of small-scale reindeer herding, a prospect that has prompted recent discussions of introducing quotas.

In Norway, the system of support (financial support or subsidies) for slaughter or meat production has recently been revised to deal with the problem of large herds. Opinions on this change are divided; for instance, one actor noted that 'The Saami Parliament has itself been sceptical regarding the change, where you now make a relatively large part of the funding dependent on production, which of course means that those who have the largest herds get the largest share' (Saami Parliament, Norway).[64] Both the NRL, the major reindeer herding organization in Norway, and the responsible state department, Landbruksdepartementet, which has been promoting the change, were seen by some actors as focusing more on large-scale reindeer herders and a rationalization of reindeer herding than on small-scale practices (Saami Parliament, Norway): Indeed, as one Norwegian herder noted, 'the NRL has always been far out on the right of the political spectrum and that is the reason why we have received such production benefits, that selling is what is rewarded' (Reindeer Herder II, Norway).[65] However, one motivation for the change was to raise the incentive to limit herd sizes, as these have been rising above the state-determined limits of sustainability in some areas; lichen stands have become badly overgrazed in some areas. As the Reindeer Trade Authority in Norway noted:

> *If we cannot gain control, we have many favourable years and we cannot control the number of reindeer and, what's more, the market situation cannot take care of the production, then we will fall into the same situation again with the number of reindeer too high where you start grazing on your capital. Today we have perhaps, as I see it, already begun to do so.* (The Reindeer Trade Authority, Norway)[66]

The problem in trying to limit reindeer herd sizes is that large-scale reindeer herders resist decreases in their capital. Their reasoning is that if they agree to decrease the herd sizes, others may build up their herds in common grazing areas (The Reindeer Trade Authority, Norway). In addition, as there is no longer any target price for reindeer meat, prices may fluctuate more dramatically and herders may become reluctant to slaughter at falling prices (Stakeholder meeting, reindeer herding, Norway). The larger-scale reindeer herding in Norway may also be seen as resulting in a better reindeer herding economy and less of a dependency on keeping the slaughter outtake at a maximal level. However, one factor contributing to the high reindeer numbers in Norway may be the traditional status of having large herds among the Saami (Reindeer Herder II, Norway). There are also some perceived problems with the particular Norwegian system of husbandry units, whereby the right to obtain support for reindeer herding is granted to an administrative unit. This unit may be made up of several people but is only registered in the name of one, to whom the support is granted. The limitation on the number of available administrative units restricts recruitment into reindeer herding and gives rise to different classes of reindeer herders. 'As long as nothing is done to change the [support] distribution policy and those kinds of things, the only rational thing for a reindeer owner to do is to increase the herd' (Saami Parliament, Norway).[67] The situation can be described as a classic 'tragedy of the commons', where commons are depleted as they do not really belong to anyone and short-term maximum utilization may be the most rational choice.

To deal with this situation, state regulation on reindeer herding stipulates that if a district acquires a higher number of reindeer than allowed, no support will be received for any number of animals in excess of 600 for each husbandry unit (The Reindeer Trade Authority, Norway). The state has also attempted to restore some of the management responsibility to the local and regional levels by implementing increased self-regulation so that reindeer herders themselves can coordinate and limit any competition for grazing areas (The Reindeer Trade Authority, Norway). In some planned measures:

> *... in the district plan ... the greater district will define that here is that* siida's *[the traditional Saami cooperative herding unit] winter grazing area and that* siida's *and that* siida's *and so on. It should be made clear. If a particular* siida *thinks that it has too many reindeer, it should have the opportunity to decrease the number and benefit from that reduction by gaining a larger area per animal. That is the idea of a greater district ... to remove some of the competition.* (The Reindeer Trade Authority, Norway)[68]

One reason why the Stortinget (Parliament) opts for self-regulation may be that the practice of reindeer herding is 'purely Saami ... the Parliament ... does not dare to implement too harsh measures ... or forcible measures' (The Reindeer Trade Authority, Norway).[69] One Norwegian reindeer herder noted, however, that this may result in the problems being transferred to the regional level, instead, where there is no one to turn to for a solution (Reindeer Herder II, Norway; cf. Saami Parliament, Norway). There is thus a limit on what can be accomplished through individual efforts. With reference to changing the basis on which reindeer numbers are distributed from a district to a reindeer husbandry unit (a quota), herders noted that one risk in the new system would be that the system might be based on existing reindeer numbers; in other words 'everyone will end up with 600 [the highest reindeer number you can have in order to gain support] if a quota is introduced' (Reindeer Herder II, Norway).[70]

However, if reindeer numbers are not reduced, there is also the risk of emergency (forced) slaughter. This would have several impacts on the market: for instance, if emergency slaughter is expected, meat buyers may delay purchases in expectation of lower prices due to increased supply. This is one reason, one reindeer herder argued, why Norway should bring back the target minimum price for reindeer meat that was removed in order to accommodate free market principles (Reindeer Herder II, Norway). There is thus extensive interaction between the economic and political steering mechanisms in the reindeer herding sector, especially as regards efforts to get the most out of the allowed number of reindeer.

In Sweden, a main concern at the time interviews were conducted was the reduction in state slaughter support, which resulted, among other things, in herders slaughtering as many reindeer as possible before the date on which the slaughter support ended (Reindeer Herder II, Sweden; Reindeer Herder I, Sweden). There have also been debates in the country about the best way to distribute support, one key issue being whether a system of live herd support (such as in Finland) would be preferable (Reindeer Herder I, Sweden).

In contrast to Finland and Norway, there is less of a focus in Sweden on limiting herd sizes or raising the productivity of reindeer when herd sizes cannot be increased further. This may stem from the fact that Swedish reindeer herding (at least in the case study area) is based on natural feeding, with supplementary feeding only taking place in emergencies. Herd size is limited by this reliance on natural feeding in what is a heavily forested area and further constrained by there being relatively few herders. Another reason may be that supplementary feeding in the case study area has not been able to develop because of the cost of fodder. A Swedish reindeer herder noted that:

> *The reindeer number ... sort of limits itself ... that is, there is a limited grazing area, it is not possible to have an unlimited number of reindeer, that's the simple fact. You could increase reindeer numbers if there were no upper limit, but there is, and with grazing areas shrinking rather than increasing, that limit is going to drop.* (Reindeer Herder I, Sweden)[71]

Similarly, 'the grazing area – that is what determines the reindeer number ... it is the grazing area and the grazing area only that determines the number, because [overgrazing] backfires; if you overgraze, it comes right back at you' (Reindeer Policy, Sweden; cf. Reindeer Herder III, Sweden).[72] The main way in which competition was discussed, then, was with an awareness that the number of herders should not be so large as to prevent economic viability: 'If you as a company are going to find a level where you can make a living, you cannot have an unlimited number of companies. And at the same time, you cannot have too few if you are going to practise effective reindeer husbandry' (Reindeer Herder I, Sweden).[73] The main focus where competition is concerned in reindeer herding in Sweden is on additional income from other sectors, such as tourism, rather than on reindeer numbers.

4.3.2. Impacts on reindeer herding from multiple use of forests

Actors in reindeer herding work within the political-economic system to adapt to the sector's poor profitability, which is limiting herders' possibilities to continue their livelihood. Outside this system, however, there exist additional constraints on the livelihood that originate with other sectors and actors that use the same areas.[74] The importance of such external impacts was emphasized in all three case study areas. This concern is reflected in the following comment: 'If you view reindeer herding in comparison with other industries, you see a great deal of external pressure on reindeer herding. That is one of the main threats to reindeer herding today' (The Reindeer Trade Authority, Norway).[75] These conflicts are typical of a multi-user situation: 'When it is a question of the same resources and who can use what and how, the fact is that there are these tensions between different sources of livelihood' (Reindeer Herders' Association, Finland).[76] Reindeer herders thus emphasized that herding must be seen and evaluated in terms of its land use and disturbances caused by all of the other actors that use the same areas, disrupt herding and encroach on pasture land; they may even make some grazing areas unusable and thereby contribute to overgrazing on the remaining pastures (Reindeer Herders' Association, Finland).

The sectors emphasized as impacting reindeer herding were, depending on the area involved, forestry, tourism and recreation; other factors cited were infrastructure such as roads, summer home construction and wind power. Predation and societal support for maintaining predator populations was also discussed as a form of encroachment, while environmental protection in general was seen as having both positive and negative impacts.

In both Finland and Sweden, the main source of encroachments on reindeer herding is forestry: 'For reindeer herding the biggest question is, well, that question of pastures, and they want to have restrictions on forestry. It's a basic question' (Lapland Regional Council, Finland; cf. Reindeer Herder III, Sweden).[77] The main problem here is that forestry logs old forest where arboreal lichens grow, depleting emergency fodder for reindeer and forcing increased supplementary feeding. Soil preparation and other silvicultural measures also limit the area that can be used for reindeer herding (Reindeer Herder I, Sweden; Reindeer Herder II, Sweden); this is a problem for which some actors think forestry should compensate reindeer herding (Stakeholder meeting, reindeer herding, Sweden). These problems have existed for a long time and have led to agreements and practices involving consultation in both Sweden and Finland. In Sweden, for instance, bodies exist on multiple levels that enable forestry to consult with reindeer herding, on both a regulatory and a voluntary basis (such as, more recently, through forest certification, which is a voluntary market-based labelling system in which individual forestry units agree to follow certain ecological and social principles for logging). Many problems remain, however, and reindeer herders in general feel that the livelihood receives limited consideration by forestry, that forestry has economic impacts on their livelihood (by removing grazing areas that herders then have to replace with supplementary feeding) and that they have limited opportunities to influence matters. Forestry and reindeer herding both see a need for improving each sector's knowledge of the other.[78] An impediment to this so far, as well as to actions that might reduce forestry impacts on reindeer herding, has been the limited documentation of the effects of different forestry measures (such as soil preparation). Accordingly, actions have been taken, for instance in Sweden, to develop planning for reindeer herding similar to that used in forestry, to be used in consultation, and to develop education on consultation for both groups (Reindeer Herder I, Sweden; Forestry and Reindeer Herding Administration, Sweden; Reindeer Herder II, Sweden; Reindeer Herder II, Finland; County Forestry Board, Sweden).

While forestry is the party that is the main source of potential friction in Sweden and Finland, considerable attention was focused in Norway on infrastructure, especially wind power development (although wind power was mentioned elsewhere, cf. Reindeer Herder I, Sweden). For instance, one Norwegian reindeer herder noted that although they have no wind power developments in his area as yet, there have been several applications, some for large-scale developments. This is in line with the Norwegian policy focus on green electric power and subsidies for wind power development. If this policy were to be implemented, 'we estimate that half of the summer grazing grounds would be affected (Reindeer Herder II, Norway).[79] As a result, the same herder noted, 'We have had to hire a lawyer because we cannot do this ourselves. We would not have a chance because the actors are so large. There is state power, there is environmental power ... [Wind power] is the largest problem we have' (Reindeer Herder II, Norway).[80]

Roads are the chief concern where infrastructure is concerned in Sweden and Finland, as they provide access for numerous actors. Given that reindeer herding also uses the roads and that road construction is largely completed or has been curtailed for economic reasons, it is not construction but the use of roads that has figured more prominently in the discussion. In the case study area in Sweden, the car testing sector, which has expanded in the area recently, has in some cases impinged on reindeer herding (Reindeer Herder II, Sweden). Another infrastructural problem is train tracks in the areas, which sometimes affect herding. There is a potential for accidents both for reindeer and herders where reindeer cross the tracks, and an additional stress in that animals may go into areas that are not properly fenced (Reindeer Herder III, Sweden).

Tourism and recreation are sometimes perceived as more beneficial than other sectors as they represent one direction towards which reindeer herding may diversify. Tourist travel by snowmobile or dog teams may, however, disturb reindeer. In Norway, four-wheeled bikes (ATVs) are a particular problem in summer, and the possibilities to limit the number of special permits for these are a matter of discussion. Tourism in the form of infrastructure development and accessibility is otherwise of lesser importance in the Norwegian case study area because of the relatively stringent legislation on areal development in Norway: 'On the Finnish side you have small camping places along the entire river, but on the Norwegian side you have [one] camping site' (Reindeer Herder II, Norway).[81] On the other hand, 'tourism is very much up and coming. It is a future impact' (Reindeer Herder II, Norway).[82]

Predation is a common problem in all countries, although views differ as to how severe a problem it is between different areas. One principal problem here is that herders feel they are not fully compensated for the reindeer lost to predators (Reindeer Herders' Association, Finland). This perception arises because herders only receive compensation for a particular number of animals lost or if the body of an animal killed by predators cannot be found.

Finally, environmental protection is seen as having both positive and negative impacts on reindeer herding. These depend on whether the areas set aside for environmental protection are available for use in reindeer herding and whether their use is coordinated with reindeer herding needs, which is not commonly the case. With the increasing environmental protection, there is a worry, especially in Finland, that reindeer herding will not be allowed in national parks (Reindeer Herder II, Finland). There are differences in views as to whether reindeer herding should be seen as a 'natural' practice that is allowed in conservation areas or not. Some differences can be observed between countries on this question: in Norway, reindeer are seen as wild animals which are allowed in reserves; in Russia, on the other hand, they are considered domestic animals and are not allowed in reserves. 'So when a Norwegian reindeer walks over to the [Russian side of] the reserve, the Russians cannot see … how reindeer can graze there' (The Reindeer Trade Authority, Norway).[83] In Sweden, reindeer are allowed in reserves, and the development of nature reserves thus offers some support for reindeer herding. On the other hand, areas set aside for nature reserves reduce the amount of land that can be set aside for reindeer herding, and the areas considered important for reindeer herding and conservation are seldom the same. At a Swedish County Forestry Board, one interviewee noted: 'The biggest problem today as I see it is that the County Administrative Board that governs nature protection and reindeer herding is not discussing protection … if you allocate several hundred million SEK for reserves, it would be desirable … to combine [the two interests]' (County Forestry Board, Sweden).[84]

Major problems for reindeer herding include coordination with a variety of actors, something that is seldom accomplished through national regulation. Coordination places considerable requirements on both reindeer herding and regulation, as herding often does not have resources for additional administration. A substantial effort would be required here for the state to carry out the necessary coordination with multiple actors and to establish networks within ongoing processes such as environmental protection. One partial solution to the conflicts may lie in better coordination between actors, even if it is unclear how this should take place:

> *When you also live in such a small community you do not in any way want to halt development ... we do not want to live here all alone; we are as dependent on infrastructure and the society at large as everyone else. But I am convinced that one can do things much better by perhaps starting conversations at an early stage with the Saami village in question, find suitable areas and perhaps cooperate to a larger extent than has been the case.* (Reindeer Policy, Sweden)[85]

The problems for reindeer herders stem from the fact that the livelihood both requires a great deal of land and is a relatively small sector economically. Some actors in Norway also noted that the focus on rationalization in reindeer herding may result in further problems in the future regarding competition with other land uses:

> *If the number of people working in reindeer herding gets very small, in other words you rationalize to reduce the number of people in the business, the importance of herding for Saami culture will decrease and the argument for having reindeer grazing lands and for using large areas for reindeer herding will become weaker. Thus there is no basis for, what should I call it, setting aside almost all of Finnmark [for reindeer herding], since there are very few people who make a living from it.* (Saami Parliament, Norway)[86]

Such a development would thus undermine the arguments for the special cultural standing of reindeer herding as a reason for its land use.

4.3.3. International norms and regulation impacting reindeer herding

Reindeer herding is largely a livelihood of limited economic standing, one that often competes with larger land-use sectors. Herding has thus far had only limited possibilities to influence these relationships or state regulation. Currently, however, questions regarding the conditions for reindeer herding are greatly influenced by ILO Convention No 169, an international instrument concerning the right of indigenous peoples to land and water, which Norway was the first party to ratify, in 1990. Ratification of the convention was being discussed increasingly at the time of the study in Sweden and Finland, to some extent as a result of pressure on these countries from the UN and the EU. Consequently, the convention and its possible impacts on the areas figured prominently in discussions with interviewees in the Swedish case study area, where the convention has been used as an argument for demanding more extensive rights for reindeer herding and has thus raised some concerns, especially in forestry. 'The ILO ... brought this question [about rights] into an entirely different perspective than one originally had about regarding the Saami as an indigenous population' (Forestry and Reindeer Herding Administration, Sweden).[87] In Finland, the interviews were carried out in an area where herding is carried on primarily by non-Saami Finns, so the ILO convention was less of an issue, as it would mainly impact Saami reindeer herding further north (Common Forest I, Finland; Common Forest II, Finland). In Norway, where the convention has been signed, reindeer herders also did not discuss it much; the newspaper survey, however, revealed that the ILO convention has been used for political purposes in the case of both reindeer herding and, especially for the Saami, fishing, a sector of larger economic importance which will be discussed in the following chapter.[88]

In Sweden, where the issue was discussed most among the case study areas,[89] one interviewee whose work involved coordinating cooperation between reindeer herding and forestry noted that while reindeer herding problems have been roughly the same with regard to encroachments since the 1970s, the ILO ratification process has raised the profile and intensified discussion of the issues politically. This is to some degree due to international pressure: 'Another thing that has helped these issues get wind in their sails is that last week we received signals from the EU, and earlier from the UN too, that Sweden has not been handling this situation well' (Forestry and Reindeer Herding Administration, Sweden).[90] This impact on Sweden and also Finland, in the opinion of this interviewee, can be attributed to the standing of Saami representatives internationally:

> *[In] the UN Permanent Forum on Indigenous Issues ... the chair is [Saami representative] Ole Henrik Magga ... he deals with 5000 different indigenous peoples' groups around the world ... and that results in an enormous pressure on, well, Finnish and Swedish and Norwegian Saami and to some extent also Russian Saami.* (Forestry and Reindeer Herding Administration, Sweden)[91]

Even before any ratification, the possible national interpretation and implementation of the ILO convention has thus prompted extensive and heated discussion in both the forestry and reindeer herding sectors. The process has perhaps attracted the considerable attention it has because of the large areas and the many actors whose interests are involved:

> *What you often forget in Sweden is that the reindeer herding area is about eight million hectares of forest land ... Those who live and work in this, if you look at the forest side, that is 40,000 private user units plus large-scale forestry ... That should be compared with about 2000–2500 Saami ... There you have a bit of background to the conflict [between reindeer herding and forestry] ... there are very many private persons who are involved in this.* (Forestry and Reindeer Herding Administration, Sweden)[92]

In Sweden, the political process to evaluate an eventual signing of the convention has, among other things, included the appointment of a number of state committees to study the issue. '[A]ll of this makes [the reindeer herding question] incredibly hot right now, so in a very short period of time there have come very weighty reports and committee reports which have stirred this pot' (Forestry and Reindeer Herding Administration, Sweden).[93] Among these is the border committee (*gränsdragningskommissionen*), for which one actor noted that when it was established 'all hell broke loose ... in northern Sweden that now there will be a transfer of rights from the so-called Swedes to the Saami and what will happen and will they receive a veto concerning the use of our land and so on. Then the discussion started in earnest' (Forestry and Reindeer Herding Administration, Sweden).[94] Another committee, on reindeer herding policy, also provoked discussion with some of its proposals:

> *[The reindeer herding political committee] in principle, in some suggestions, considers that the Saami should have a 'veto right' where logging is concerned. I do not think that you should interpret it that way, but many do. And then you can read their suggestions, those of the committee on reindeer herding policy, a bit, as they say, like the Devil reading the Bible.* (Forestry and Reindeer Herding Administration, Sweden)[95]

This increasing international discussion of indigenous peoples' rights has resulted in major concerns about how forestry will be affected locally. Thus, for instance, an interviewee in forestry was concerned that the convention would be signed because of pressure from the EU and other bodies, with less regard given to its application locally:

> *The threat is that people sit at Rosenbad [the government] who do not have anything at all to do with the conflict, [or know] how it works ... I do not know if the purpose is to gain brownie points in Brussels, or why they will sign the ILO convention without making it clear to themselves what consequences it will have.* (Common Forest, Sweden)[96]

The conflict largely centres on rights to resources and the perception of ownership that different groups have regarding these. As these are determined by law and regulation, the possibilities to effect or avoid legal changes take centre stage. Thus, 'The Saami organizations want more rights but there are also other groups in Västerbotten and Norrbotten [counties] who think that they should have more ... hunters and organizations and those who claim that they have land rights' (County Administration, Sweden).[97] The same interviewee noted that both those who consider themselves Saami and those who see themselves as having other, unclear or mixed ethnic or linguistic backgrounds but who have lived in the areas for a long time often perceive themselves as having land use rights: 'In Norrbotten and Västerbotten [counties] ... there is a local population, often far-flung Saami descendants who have ceased reindeer herding and who consider themselves to have a greater right to these lands than the state has' (County Administration, Sweden).[98] When a decision on ratification will be made and ratification itself are highly political topics and seen as requiring ways to discuss resource use that do not cause local conflicts (Forestry and Reindeer Herding Administration, Sweden).

Interviewees highlighted, however, that the conflict is sectoral and that on a personal level the interactions between actors may be rather amicable. The following statement typifies the situation: 'The personal relations with forestry are quite good; it is these overarching goals that we do not agree on' (Reindeer Herder III, Sweden).[99] One indication of the large positive personal interaction across the sectors is that in the case study area in Sweden there are many non-Saami who own reindeer: perhaps 10 or 15 per cent of the herd in one of the Saami villages where interviews were conducted was owned by '*skötesrenägare*' who are not part of the administrative unit of the Saami village but pay the village for looking after the reindeer and often help out with the herding when it is time to gather or slaughter the animals (Reindeer Herder II, Sweden). There were also interviewees who both own reindeer and work in forestry (Forestry Machine Driver, Sweden; Collective Forest I, Finland). In the Finnish case study area, reindeer herders also often own forest, and therefore earn income from both forestry and reindeer herding (Reindeer Herders' Association, Finland).

There are other international developments in addition to the ILO convention that may impact the reindeer herding–forestry relationship. The revision of FSC forest certification (a voluntary agreement on ecologically and socially sustainable wood production that is widely adopted in Sweden) may add further features that support reindeer herding. This was discussed by Swedish interviewees: 'The new signals [in] the FSC [are that] you look at forest with arboreal lichen and [should] weigh [its importance] in ... the overall situation for the Saami villages when you talk about [adequate] consideration' (Forestry and Reindeer Herding Administration, Sweden).[100] There is some hope that the improved forestry–reindeer herding coordination required under the FSC scheme may increase mutual understanding among the sectors by, among other things, providing an incentive for forestry to supply resources for reindeer herding land-use planning (Reindeer Herder I, Sweden). In Finland, this issue was not taken up in the interviews – reflecting the relatively smaller impact on forestry of the less demanding PEFC certification common in the country – although it had been discussed to some extent in the press.[101]

4.3.4. The governance network perceived by reindeer herding

Reindeer herding is a sector that is impacted in its land use by many larger sectors and many levels – predominately national state regulation and support – and has few strong or well-developed networks protecting its production interests. At the same time as the requirements imposed on reindeer herding are increasing, a thinning of the sector's structure is taking place, reflected in fewer active reindeer herders and a greater reliance on technology. The social and organizational structure of reindeer herding has changed. Where people lived in close contact before, today 'the common time for those who herd together has become more limited than before and the distances between them have increased, which in fact results in a need for regulation, perhaps for something as modern as organizational development' (Saami Parliament, Norway; cf. Reindeer Research Centre, Finland).[102] Other actors also suggested that this could be accomplished by developing a management culture in the reindeer herding associations and making them into stronger interest organizations that are responsible to the reindeer herders:

> *We should have a functioning government in the reindeer herders' association [and] it should be possible to make demands of the government ... Education should be arranged for the members of the government because they do not have a management culture ... It would then be a fair game, but no such system exists here.* (Reindeer Herder II, Finland)[103]

Somewhat similarly, another actor noted that reindeer herders also need to adjust to a conflict between their own interests and common or societal pressures (Reindeer Research Centre, Finland). Such conflicts with sometimes unclear decision-making structures also limit the possibility of agreement on resource use between different actors such as forestry and reindeer herding. At the same time, the need for consultation with and knowledge of other sectors and administrative systems has increased:

> *We have coordination with tourism, forestry, county councils, municipalities, regional hunting organizations, regional fishing organizations ... we have an enormous number of such things that we have to keep up with today ... societal planning proceeds so quickly.* (Reindeer Herder III, Sweden).[104]

However, reindeer herders noted that they have relatively little to do with municipal administration outside particular consultations (Reindeer Herder II, Norway). For instance, a Norwegian reindeer herder noted:

> *To deal with four municipalities and the whole bureaucracy is hopeless and at the same time the municipalities do not even have that much decision-making power. Things go beyond the municipality in any case if*

> *there are bigger issues; so it is more the county level we deal with in things such as driving, bare ground driving, predators and the larger development projects, because the municipality is only a consultative authority and the issue goes on to the county council in any case.* (Reindeer Herder II, Norway)[105]

Some actors also noted that the regional level has very little direct impact on decision-making. For instance, the Regional Council of Lapland, Finland, notes that the its regional influence is very small, as the relevant decisions are taken by Metsähallitus (the Finnish Forest and Park Service) and the Ministry of the Environment (Lapland Regional Council, Finland; cf. Metsähallitus, planning). The same is noted in Norway: the administrative level with which reindeer herders interact is not that which makes the far-reaching decisions. For instance:

> *The Reindeer Trade Authority is the external actor for the Department of Agriculture so blaming the Reindeer Trade Authority is the same as shooting the piano player. That is, the composer is the department … it sets out the major lines with regard to reindeer herding management.* (Saami Parliament, Norway)[106]

For the most part, reindeer herders are not able to influence decisions concerning them by this interaction. That there is a perceived need for feedback on state policy from the local level, however, can be seen in an example from the Norrbotten County Administrative Board. After inviting feedback on its new organization policy, the board 'received a lot of points of view but very many concerned [forestry–reindeer herding] rights that are not the concern of the county level at all' (County Administration, Sweden).[107] It thus seems that there exists no clear channel to express concern and exert influence on broader issues. For instance, the reindeer herding unit of the county council has an opportunity to provide feedback to the Swedish Board of Agriculture through meetings, but much of the communication is informal, the understanding being that the county council is an executive rather than a decision-making authority (County Administration, Sweden).

Reindeer herding may thus be seen as a sector whose networks are mainly restricted to limited intra-sectoral cooperation, to some degree to regulation (particularly at the national level), and to the limited networks related to meat production and sales. The Finnish, Norwegian and Swedish reindeer herding administrations do not have much to do with each other, except for coordination and the work of border committees (The Reindeer Trade Authority, Norway). These relationships are illustrated in Figure 4.1, which shows the multi-level governance network for reindeer herding, in other words the actors seen as impacting reindeer herding in the local case studies (as perceived by interviewees).

Reindeer herding is here seen as relatively thinly structured, with most of the relevant organizations placed within regulative structures. The actors that local reindeer

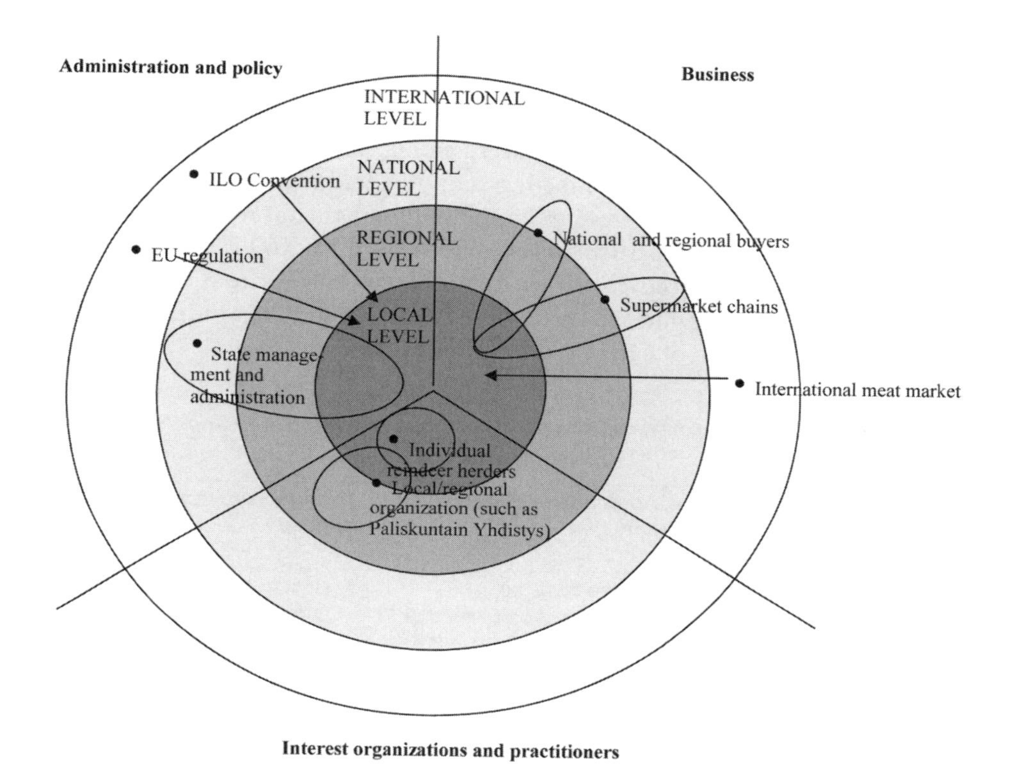

Figure 4.1 *Multi-level governance of reindeer herding*

The diagram is divided into three sectors corresponding to the main groups of actors (administration and policy, business, and interest organizations and practitioners). Four levels of organization are distinguished: local, regional, national and international. The figure shows the principal organizations active in each sector, with dots indicating the level at which each nominally operates. Ellipses reflect the scope of an organization's influence as perceived by interviewees and arrows indicate large-scale influence via the market or regulation.

herding must engage with include reindeer herding organizations (such as district organizations in Finland or the NRL in Norway), meat-processing companies, meat buyers and supermarkets, fodder sellers, local and regional government authorities, and researchers (particularly in Finland). Herders are also required to deal with national legislation and organizations such as those dealing with limitations on reindeer numbers and policy, EU support and regulations, norms on indigenous peoples as manifested in the ILO convention and in certification, and external impacts such as forestry, infrastructure, tourism and environmental protection. Reindeer herding was also not linked to employment in the way forestry was, which may be seen as further limiting its possibility to develop networks locally (possibly with the exception of the Norwegian case study area, where reindeer herding has greater economic importance than in Sweden and Finland) (Local Government I, Norway).

All in all, the regulation of reindeer herding has increased greatly during the time interviewees have worked in the livelihood, and the research reveals an increasing, rather than decreasing, role on the part of the state, possibly related to the current political importance of reindeer herding. Yet reindeer herding finds it difficult to assert itself against competing land uses, and it is here especially that international means to exert pressure on the state – the main regulatory body – have been emphasized. In particular, ILO Convention No 169 is used by actors in reindeer herding to improve their position as small-scale interests against large and economically strong interests such as forestry, which indicates the increasingly international dimension of national and regional politics.

Table 4.1 *Main perceived changes, adaptations and limits on adaptation in reindeer herding*

Changes in the governance framework (economic and political changes)	Adaptations to perceived changes	Limitations to adaptive capacity
• Increased competition over pastures (including logging of old forest) • Technological change (such as increased mechanization) • Stabilized meat production • Increased costs for livelihood (including technology and fodder) • Limited recruitment • Supplementary feeding • Increased demands on administration • Increased reliance on market with fluctuating meat price and limited market network • Competition with other meat production	• Increasing the consumption of reindeer meat through advertising, marketing and processing • Increased market cooperation between countries • Suggestions of a slaughter support system and different national systems for limiting reindeer numbers • Additional support systems (compensations and subsidies) • Adaptations to maximize herd productivity • Political pressure (for instance through the ILO) on increased resource rights • New opportunities for coordination with forestry through forest certification • Changes towards larger-scale and rationalized (rather than small-scale) systems	• Limited meat transportation and market systems • Seasonal nature and limited scale of meat production • Limited national reindeer numbers • Difficulties in implementation (e.g. of limitations in reindeer numbers) increasing competition between herders • Different legislative possibilities for reindeer herding compared, for instance, with forestry • Limited coordination between, for instance, reindeer herding and environmental protection aims for land • Limited knowledge of reindeer herding among decision-makers, and limited access for reindeer herding to decision-makers

4.4. Climate change

Climate change constitutes an additional stressor on reindeer herding that particularly impacts natural grazing areas and grazing conditions. Climate and weather conditions impact reindeer herding to quite a large extent. Unreliable or unfavourable weather conditions may increase the need for emergency feeding and thereby increase costs. Favourable weather may result in an oversupply of reindeer meat, as has been seen in the cases described above. Because reindeer herders need to adjust to the weather day by day in some cases, they are aware of climate and weather and the impact that even slight long-term changes in conditions or the occurrence of extreme weather events could have: 'The weather conditions are a critical factor for the wellbeing of reindeer' (Reindeer Herders' Association, Finland).[108] Variability is here a problem in itself: 'For reindeer herding it is best that summer is summer and winter is winter. It is that straightforward' (Reindeer Herder II, Norway).[109]

Especially because of the warm trend in the last few years, many of the interviewees were aware of weather changes and the risks specific changes can bring. In the three years preceding the study, conditions were favourable, because there was little snow, early snowmelts and warm summers with good grazing (Reindeer Research Centre, Finland; Reindeer Herder II, Norway). The situation was thus relatively close to ideal: 'Easy winters, little snow in winter, an early spring and no mosquitoes' (Reindeer Herders' Association, Finland).[110]

A period of some ten years preceding these three good years, however, was marked by comparatively unfavourable weather conditions, which were especially bad around 1990 (Reindeer Herder II, Norway; Reindeer Herder I, Norway). At that time, 'there was mild weather in autumn, early in autumn, and then it froze and that was bad' (Reindeer Herder I, Norway).[111] However, some herders noted that the recent warm weather may represent a deviation from previous (or 'normal') fluctuations:

> *The way it has been now, such swings in temperature that happen so fast – we have minus 20 degrees in the morning and then in the afternoon we have plus 2 – everyone says that it has not been that unstable before ... This happens up until Christmas and in recent years we have had no snow until Christmas ... People think that is really strange.* (Reindeer Herder II, Norway)[112]

Similarly, 'I do not remember that when I started [about 30 years ago] we had these thaws in the middle of the winter ... I feel that we had a somewhat more stable climate' (Reindeer Herder III, Sweden; cf. Reindeer Herders' Association, Finland).[113] Conditions such as winter thaws could, among other things, result in problems for actors who cannot afford larger-scale supplementary feeding in that they would have to go farther 'up into the mountains all too early and end up with much worse production and then the profitability drops although you have high costs' (Reindeer Herder III, Sweden).[114]

What is more difficult, however, is to say that the warmer years that have prevailed recently are part of a trend that is the result of climate change: 'One may say that the temperatures have been milder, but ... it is a short time period that is being discussed' (Reindeer Herder I, Norway).[115] Thus, for instance, a Finnish local government official noted that while there have been warm summers recently, there have also been some extremely cold weeks in winter: 'There has been talk about global warming, that the climate begins to warm in the worst case, but two or three years ago we had almost –50°C for over a week – it was something like 45–48 degrees below zero' (Local Government, Finland).[116] It is also difficult to make long-term conclusions based on small changes (Reindeer Herders' Association, Finland), and difficult to predict climate change (Reindeer Research Centre, Finland).

The following sections will discuss the risks or opportunities perceived for the autumn and winter and spring and summer seasons. The final sections of the chapter will then discuss possible adaptations to changes overall, including changes in climate.

4.4.1. Autumn and winter

Autumn is an important time for reindeer herding in all the regions, as the weather conditions then determine the quality of grazing for much of the winter. A long and warm autumn, when reindeer can graze well before it snows, is the most favourable. It is also best if the ground freezes before snow falls, after which reindeer are forced to dig for lichen. This is because 'the kind of winter you have depends on the ground, in other words how suddenly the ground freezes and how much snow there will be and of course whether you get layers of ice in the snow' (Reindeer Research Centre, Finland).[117] For instance, if it thaws or rains in winter, an ice layer may form on the snow or, in the worst case, directly on the ground (Reindeer Herder I, Sweden; Reindeer Research Centre, Finland). Relatively small differences in temperature during this time can thus have a large impact on grazing. Ideal conditions in this period are that 'it freezes before it snows, then the grazing will usually be good. And then there cannot be any mild weather at that time ... In that case, the ground won't thaw and then the grazing is absolutely superb and if snow also falls on the ground it is very good' (Reindeer Herder II, Sweden).[118] At the start of winter, then, there should be 'dry weather until October, preferably in the end of October I would like it to start snowing' (Reindeer Herder III, Sweden).[119]

However, it is important that snow does not come too early, as this causes reindeer to lose weight. If snow comes relatively late, the reindeer will 'have time to eat their fill after the rut' (Reindeer Herder II, Sweden).[120] If it snows late, this increases the period during which reindeer can graze on summer forage such as green plants and grasses, meaning that they do not need to spread out over large areas to find food. Herders then need less fuel and can keep petrol costs down because they only need gather the animals from a smaller area (Reindeer Herder I, Finland; Meat Buyer, Finland). Too little snow is not beneficial for herd management, even though it is good for grazing: 'Gathering reindeer for slaughter and separation when there is little snow – that has been costly' and herders may then use helicopters instead of snowmobiles (Reindeer Herder II, Sweden; cf. Reindeer Herder III, Sweden).[121] Also, if there is little snow for a long time in autumn, watercourses freeze and will carry the weight of reindeer but not of all-terrain vehicles or snowmobiles. Reindeer may then disperse over large areas, with herders having little possibility to gather them (Reindeer Herder III, Sweden).

When snow has fallen, a stable winter climate is preferable: 'The ideal climate [is] about a maximum of 50 centimetres of snow in winter and a stable winter climate, that is, 10 to 20 degrees below zero' (Reindeer Herder III, Sweden; Reindeer Herder II, Sweden).[122] Snow depth is important in keeping the reindeer together and ensuring good grazing: 'There shouldn't be too much snow ... if there is over one metre of snow ... they stop digging' for lichen and may wander in search of better grazing areas (Reindeer Herder II, Sweden).[123] If none are available, supplementary feeding may be required (Reindeer Herder III, Sweden). Additionally, if there is much snow, and hard snow that will carry smaller animals but not reindeer, predators such as the wolverine or wolf may be able to track and kill reindeer, as they cannot readily escape (Reindeer Herder I, Sweden). The temperature being somewhat colder than ideal, however, is not a big problem: 'The cold does not do anything ... if [the reindeer] only have good grazing and are well fed' (Reindeer Herder I, Sweden).[124] The largest problem for reindeer in winter is if it rains and then freezes over: 'The grazing areas can become locked and there can be ice covering the entire heath' (Reindeer Slaughter House, Norway).[125] As reindeer largely subsist on ground lichen in winter, which they dig up through snow and layers of ice, any weather events that impact access to lichen are crucial. Indeed, as one herder observed: 'The worst is if the weather becomes mild and shifts between ... rain and sub-zero weather' (Reindeer Herder I, Norway; cf. Reindeer Herder I, Sweden).[126] Especially in Norway, where supplementary feeding is not so common, such weather may have extreme effects, as freezing rain prevents reindeer from lichen grazing: 'The reindeer die like flies and then there are suddenly too few reindeer. That may happen and it is nature that determines it' (Reindeer Slaughter House, Norway).[127] Winter thaws, however, can have more or less serious effects depending on snow depth and the extent of the thaw. The ice layer on the snow, if it is not too thick, will loosen up if more snow falls on it, making it possible for reindeer to dig again (Reindeer Herder II, Sweden; cf. Reindeer Herder II, Finland).

In recent years, interviewees had seen changes at this time of year that indicated climate change might in fact be under way. Possible changes in weather that interviewees noted include 'more and more wet autumns ... there is partly mild partly cold … and the grazing goes bad … that is more of a rule than an exception' (Reindeer Herder I, Sweden).[128] Interviewees in general also note the recent shorter, warmer winters. 'These last years there have been changes, very obvious changes … extreme thaws in winter … when the climate should be stable – in January or February it should not go up to seven, eight, ten degrees in the sun' (Reindeer Herder III, Sweden).[129] Thus 'last year for instance we had almost no snow at all in the month of March, something that has never ever happened' (Reindeer Herder III, Sweden).[130] Mild weather has thus been noted in the case of both the autumn and spring transition periods (Reindeer Herder II, Sweden). The changes and unpredictability – especially in terms of changes across the freeze–thaw threshold – were seen as potential risks: '[This] winter it almost destroyed the grazing completely, this thaw that we had just a week ago, so we are very concerned about weather changes' (Reindeer Herder III, Sweden).[131] A warmer climate, with more frequent fluctuations across the freeze–thaw line, would be detrimental to reindeer herding – unless it meant that winter became so short that the importance of conditions during the period diminished: 'There is a limit there: if the winter became so short that you can deal with this, maybe only three months, but I doubt we'll see that' (Reindeer Herder I, Sweden).[132] However, 'if we get more ice on the grazing areas than we have had before, if it becomes warmer and is not just a passing thing, it could change the way reindeer herding is practised' (Reindeer Herder II, Norway).[133] This could mean, for example, that 'either [herding] will become based more on supplementary feeding, or we might have to decrease the reindeer number' (Reindeer Herder II, Norway).[134] Small changes in timing and temperature across the freeze–thaw threshold could thus have a large impact on reindeer herding. For example, if it rained in the middle of winter, 'it would no longer be feasible to have a viable reindeer husbandry' (Reindeer Herder III, Sweden).[135]

Snow depth, however, does not indicate any general direction of change; it has varied across the areas. Interviewees described rather meagre snowfalls in Sweden and Norway, and cases of both much and little snow in Finland (Reindeer Herder II, Sweden; Reindeer Herder I, Norway; Local Government, Finland; Reindeer Research Centre, Finland). The conditions may differ considerably across the areas, to some extent because of specific local conditions. For instance, one Norwegian reindeer herder noted that in a coastal climate, 'When there is mild weather in winter and they get better grazing conditions [inland] … it ices over here where we are' (Reindeer Herder II, Norway).[136] For these coastal herders, 'what has been good in the last two years is that [the icing over of grazing areas] has not taken place until the first weeks of February and then it is not so long until it starts to warm up again on the ground and the sun comes out and you get bare spots in the snow' (Reindeer Herder II, Norway; cf. The Reindeer Trade Authority, Norway).[137] For some reindeer herders, the recent conditions have thus been favourable, and for others unfavourable. Those who have benefited have done so because of good grazing up until the start of a relatively late winter, access to otherwise inaccessible areas due to relatively scanty snowfall and mild weather at times when it has not caused much damage, for example, during mid or late winter.

4.4.2. Spring and summer

The transition between winter and spring is like that between autumn and winter. It is preferable that it occurs early in order to shorten the winter and make summer grazing grounds accessible (Reindeer Research Centre, Finland): 'It is positive if [the snow] melts earlier. The earlier spring [comes], the longer the green grazing season becomes, and the stronger and finer the reindeer you get' (Reindeer Herder I, Sweden).[138] However, the transition must also be predictable so that herders can plan for it. When spring melting has begun, it should continue and preferably be relatively quick (Reindeer Herder III, Sweden). The ideal spring, in one case study area, would be one where:

> *it does not start to thaw too early [but rather] starts to thaw in the month of April ... and then it freezes up during the nights and in the daytime the grazing areas thaw, and then when you get into the month of May then it can thaw night and day ... and become bare as soon as possible* (Reindeer Herder III, Sweden).[139]

The reason why this interviewee felt that spring nights should be cold is that he drives his reindeer on foot, walking with them rather than moving them by lorry. If it freezes by night, 'we can drive the reindeer ahead of us during the nights and then let them graze during the days. Then it is easy for the reindeer to walk and you can feed them during the day ... it is tiring for the pregnant reindeer and for the calves ... who have to walk in heavy ... wet snow' (Reindeer Herder III, Sweden).[140]

Impacts on the individual reindeer herder depend on the natural conditions, geography and accessibility constraints in their areas as well as on the time at which they move their animals (some start early on the year, others just before summer, depending on weather and snow conditions) (Reindeer Herder III, Sweden). In the last few warm years, it has been possible to start migration earlier, as thawing has started earlier (Reindeer Herder II, Sweden). Herders who use rivers for migration are, however, affected by the impact of the weather on river ice (Reindeer Herder III, Sweden). If there is a great deal of snow in the forest in spring, it impedes movement of the herds and the herder's group has to drive its reindeer on the river. Earlier warming may cause problems: 'Sometimes with these weather types we have had ... the ice ... has been [so] weak that herds have fallen through into the river at times and [we have] even fallen in with the snowmobile' (Reindeer Herder III, Sweden).[141] If there is a lot of snow and it gets warm early on the river, 'you are forced to truck [the animals]' (Reindeer Herder III, Sweden).[142] Whether this is considered costly or not depends on whether the herder has help in making the migration (for instance within a large herding unit): 'It costs [money to truck animals] but for those herding units that have only a few people working in them … it may not make a large difference whether they truck them or not' (Reindeer Herder III, Sweden).[143]

Another reason why recent years have been seen as so favourable is that the warm springs have not been interrupted by cold spells with, for instance, wet snow late in spring, which may damage or kill calves (Meat Buyer, Finland; Reindeer Herder II, Sweden; Reindeer Herder III, Sweden). In some areas, such problem conditions have not occurred for a long time; in others, such events have taken place relatively recently. One Swedish herder noted: 'Since I started, it has never happened [that we have got wet snow that has killed calves] … it seems as if summer comes directly and now there are such extreme, warm summers' (Reindeer Herder II, Sweden).[144] In the Finnish case study area, on the other hand, one interviewee recalled a year in the beginning of the 1990s when snow was still falling when the calves were born: 'It was still snowing horribly and they died' (Reindeer Herders' Association, Finland).[145] Illustrating the variability in weather, the same interviewee noted that 'in less than ten years the weather has reached two extremes. This is an extremely interesting phenomenon' (Reindeer Herders' Association, Finland).[146]

In summer, then, the ideal is a warm but not too warm summer, where grazing areas grow well but water and small creeks do not dry up. It should not be too hot for reindeer to stand and graze. Optimally, there should also not be any mosquitoes or other flies that disturb grazing (Reindeer Research Centre, Finland). 'Ideally, it should be suitably warm for a number of weeks during the summer, and cold at night, and raining once a week or so, so that new grass comes up, and then August is cool' (Reindeer Herder III, Sweden).[147]

In the Swedish case study area, recent summers were described as almost too hot: 'The summer should be suitably warm, but it cannot be too warm: we can't mark the calves then, because the heat takes such a toll on them' (Reindeer Herder III, Sweden).[148] If it is too warm, 'they will not graze at all, there are mosquitoes … and different deerflies irritating them all the time and driving them from these fertile areas where there is a lot of grass and shrubs' (Reindeer Herder III, Sweden).[149] During such a heat wave some summers ago, 'our salvation was that it was so dry that the mosquitoes did not hatch' (Reindeer Herder III, Sweden).[150] On the other hand, such dry weather 'is not good for the reindeer because the snow patches dry up and even the grass dries up' (Reindeer Herder III, Sweden).[151] In these recent warm years, reindeer have stood in open areas to avoid mosquitoes and the heat in the valleys, but as there is little water there and grass may dry up, 'they do not get the growth that they require' (Reindeer Herder III, Sweden).[152] However, during one recent summer there were some beneficial factors that alleviated these problems: 'Then we were lucky, in the beginning of August … there came rain … there were a lot of mushrooms [that the reindeer could eat]' (Reindeer Herder II, Sweden).[153]

During the recent warm summers small watercourses dried up: 'We have seen that these summers, there has been very little water ... the water flow is not normal' (Reindeer Herder III, Sweden).[154] While this affected access to water by reindeer, it also meant that it was easier for the animals to cross watercourses and mix with animals from other herds. This in turn required herders to set up closer supervision, costing them more in petrol and time (when they could have been mending fences and so on) (Reindeer Herder III, Sweden). To some extent, logging has also decreased the possibilities for reindeer to find shade during warm summers: 'It mustn't be warmer [in the summer] because they have logged all these spruce forests' (Reindeer Herder I, Sweden).[155] In sum, one sees very complex interactions that determine the impact of environmental conditions on reindeer herding.

For the Finnish and Norwegian case study areas, however, the summer weather in the last few years had largely been favourable. Reindeer herders in Finland noted that summer grazing and thus reindeer growth had been good, as it had been warm and dry and there were few mosquitoes (Reindeer Herder I, Finland; Meat Processing, Finland). The reindeer had therefore been able to use areas that they otherwise could not because of mosquitoes. Dryness and the drying up of springs is not a problem in Norway, which contributes to the favourable situation of reindeer herding in the case study area: 'We live so close to the sea ... that it is always moist here' (Reindeer Herder I, Norway).[156] The warmer weather has thus been mainly beneficial in Norway. In the Norwegian case study area, there are also generally fewer problems with mosquitoes because of the coastal conditions; however, mosquito problems are more severe when they do occur: 'Our reindeer

Table 4.2 *Main emphasized climate changes and adaptations*

Projected climatic change	Local sensitivities and adaptations
Changes across the freeze–thaw line, especially causing milder winters with possible thaws	Dependent on micro-conditions such as how suddenly the ground freezes; whether the ground is dry at this point; how much snow there is when thawing occurs or whether ice is formed directly on or close to the ground; the length of the thawing period and when thawing periods occur during winter; and whether migration disruptions are caused. Increased thaws during winter may lock grazing, while a much shorter duration of winter would be beneficial
Changes in precipitation during winter, causing variation in snow depth	Too much snow makes it difficult for reindeer to dig for lichen and increases the accessibility to the herd for predators; too little may make reindeer scatter across areas, resulting in costly gathering or supplementary feeding
Warmer summers	Warm and dry summers with few mosquitoes but not so dry as to cause springs and grazing areas dry up are favourable, but may potentially result in an oversupply of meat
Later autumn	Summer grazing for a longer period is favourable
Changes in lichen and access to other grazing resources over time	System could move towards more supplementary feeding or require limitations of reindeer numbers

go into a panic … There can be specific weather conditions during a short period when there are mosquitoes and they run like mad up onto the highest [mountain] tops' (Reindeer Herder II, Norway).[157]

All in all, climate and weather conditions are important to reindeer herders, who have a very close relationship with the environment in that they need to respond to day-to-day changes in weather and are greatly affected by even relatively slight changes. For instance, the increase in the reindeer meat supply over the last few years before the study may be attributed to favourable environmental factors, which would have translated into economic benefits for reindeer herders if market networks to handle increased production had existed. This emphasizes the interlinked nature of socio-environmental systems.

4.5. Conclusion: Vulnerability and adaptive capacity in reindeer herding

Adaptation to change for reindeer herding is largely dependent on multiple factors, such as the specific grazing area and how it is impacted and the overall economic situation for the individual herder. The factors determining the economic viability of reindeer herding combine to constitute very complex circumstances, which are in turn interlinked with support systems and the limit on reindeer numbers that prevent large increases in reindeer herds (Reindeer Herders' Association, Finland). Reindeer herders in Norway noted that they may have suffered the least in the recent price fall. A variety of reasons can be cited: reindeer herding in Finnmark, relatively centralized and featuring large herds, has stronger import protection than the livelihood has in Sweden and Finland; herders use relatively little supplementary feeding and thus have lower feeding costs; the livelihood enjoys relatively favourable support systems and what some herders see as a generally higher standard of living and is comparatively less vulnerable. Herders in Norway noted that while any person living in a reindeer herding area in Finland has the right to herd reindeer, Norway has a system of husbandry units, with fewer reindeer herders acquiring herding rights but those who do having a better financial situation (Stakeholder meeting, reindeer herding, Norway). There are thus differences between the countries depending on their underlying situations and support systems understood in a broad sense.

Asked about their abilities to adapt to change, actors in reindeer herding described in both fatalistic and very practical terms adjustments that would impact reindeer herding both in their case and the cases of others in the network. Asked about their abilities to adapt to changes, actors emphasized that if reindeer herding is to adapt at all, the operating conditions for the sector must be improved. Economic factors figured prominently in this assessment.

4.5.1. Adaptations at the individual level to economic changes

The reindeer herding economy is limited especially because of low meat prices and the market situation at large, including the limited buyer structure. The restrictions on adaptation include limited funding to invest in marketing and transport, regulations limiting the number of reindeer, the very small economic niche nationally, and concrete concerns over economic globalization, this last consideration seen in the competition from New Zealand deer meat (Reindeer Herders' Association, Finland). Adaptations within these circumstances are continuous and the current situation can be seen as one of adaptation within existing means. There are several adaptations, described especially in connection with supplementary feeding and the meat market, that have served to maximize profitability in the sector and stabilize its economic situation, which is ultimately the context in which impacts will be felt.

Individual reindeer herders adapt for the most part by making reindeer herding more effective through technology and, for instance, by maintaining only small winter herds or using supplementary feeding. Adaptations to lessen the dependency of reindeer herding on unpredictable natural conditions have led towards not only increased supplementary feeding but also reindeer farming in some cases. Reindeer herders may also increase their income by selling meat themselves locally or taking on supplementary work outside herding. Today, the meat market is a principal focus and problem among interviewees, as described previously, and initiatives include increased marketing and other consumer-related actions, as well as supplementary feeding to maximize yield. The possibilities and preparedness for supplementary feeding, however, differ greatly by area depending on factors such as tradition and the cost of fodder. Other factors that constrain adaptations and adaptive capacity include the logging of old forest areas, which supported arboreal lichen, the main natural emergency fodder in both Finland and Sweden. As a result, the scope of direct individual adaptation has largely become limited to supplementary feeding. As a reindeer herder in Sweden noted, one could argue that forestry should compensate reindeer herding (Reindeer Herder II, Sweden). Supplementary feeding is thus a multifaceted strategy whose suitability varies for different actors and that may represent one choice on a path of choices that further narrows the scope of adaptation for reindeer herding. For instance, in Finland it was noted that supplementary feeding improved preparedness for unfavourable weather conditions following failures in access to grazing in 1996–1997 (Employment and Economic Development Centre, Finland). Today, however, herders increasingly rely on supplementary feeding, transforming the practice of reindeer herding and making it dependent on fodder supplies.

Rather than suggesting adaptations, however, in some cases actors stated the inevitability of the situation: when something happens, they simply have to find a way to adapt, although they cannot necessarily foresee specific adaptations now. For example, one actor stated that '[reindeer herding] has to adapt or it will die' (Reindeer Herder II, Finland).[158] The Reindeer Herders' Association in Finland described reindeer herders as fatalists, saying, 'people in the reindeer herding business have always said good times will be followed by bad times and bad times by good times' (Reindeer Herders' Association, Finland).[159] The impetus for such comments may be that reindeer herding is already adapting to its limits in some areas and feels that it cannot influence its own situation further; in such a situation, despondency over what one cannot change may be an understandable reaction.

Clear market adaptations can be seen in the suggestions for increased marketing and developing larger buyer and export networks. While these adaptations are beyond the capacity of the individual herder, some attempts been made to develop reindeer products in cooperation with supermarkets and to improve marketing in order to increase consumption. Not surprisingly, market-related and political adaptations were often seen as interconnected, especially in the situation at the time, where there was a need to create a functioning market for reindeer meat as well as to limit the number of animals being raised. Accordingly, adaptive capacity is dependent on the supply of meat, which is where economic adaptations interconnect with climate-related or environmental adaptations. Should the meat market not provide incentives to limit reindeer numbers to the allowed levels, 'then the state has to enter to perhaps pay to lower the price and increase consumption' (Reindeer Slaughter House, Norway).[160] One potential strategy could be a subsidy for customers (Reindeer Slaughter House, Norway).

4.5.2. Adaptation within the governance and political regulatory framework

It has already been noted that different impacts were mainly assessed by stakeholders in terms of their economic effects. This also holds true for adaptations within a wider framework, which were assessed and put forward in terms of how they would impact economic competitiveness, for instance. Such adaptations include lowering fuel prices to limit production costs and thus pressures on producers, raising support levels in reindeer herding to the same level as in other meat production, and paying compensation for predation (Reindeer Herders' Association, Finland). The political framework is thus perceived as crucial in that it regulates which individual adaptations are the most viable. State political actions and subsidies directly impact economic conditions. One actor noted, for instance, that 'when the state slaughter support is decreased, you have to save on other things, for example by not buying a new snowmobile' (Reindeer Herder II, Sweden).[161] The degree to which support was seen as a priority varied, however. The most common opinion was that voiced by an actor who stated that the degree to which reindeer herding can adapt individually is limited and that it is the structural policy that needs to be improved if reindeer herding is to remain viable: 'The reindeer herders have done ... a lot themselves [but an] imbalance exists in the prevailing structures, which makes it easy for us to go ahead and ask the rest of society to do something positive about these things' (Reindeer Herders' Association, Finland).[162]

Thus, for instance, reindeer herders discussed differing changes in the support systems for reindeer herding and the possibility of changing over to a quota system to modify the regulatory structure. There are many interacting factors in developing a support system that would provide incentives to keep reindeer numbers to the limits of grazing capacity. The state uses political means to implement regulations: if there are too many reindeer in a district, then 'the reindeer owners who do not follow the rules will not gain state support' (Reindeer Slaughter House, Norway).[163] Another, more drastic means is mandatory slaughter, which, given the very dramatic increase in reindeer numbers, has been discussed in Norway. Here, there is no clear political solution that could be enforced without possibly increasing conflict in the reindeer herding area.

Given the increased reliance on supplementary feeding, however, one question is how long restrictions based solely on natural grazing will remain viable, as current adaptations are increasingly detaching reindeer herding from nature. The political task of setting out a regulatory framework for reindeer herding is complicated by the fact that competing land uses, such as forestry and the transportation infrastructure, also have many impacts on reindeer herding. It is here that actors both within and outside the state are invoking international norms on indigenous peoples' rights in order to influence regulation, further complicating the situation with regard to multiple use of forests in these regions and heightening the potential for local conflict on the issue.

4.5.3. Adaptation to climate change

Climate conditions can be seen as fundamentally impacting stakeholders economically in the additional costs that are required for adaptation to the conditions or in the benefits that they provide (for example good grazing in summer, which increases reindeer growth and thus meat production). Actors assessed potential impacts of climate change as being either beneficial or detrimental to production. Autumn rains and winter thaws could reduce access to lichen stands and require supplementary feeding. For some actors, early melting of ice on rivers may result in their having to transport the herd up to the mountains in lorries or drive the animals through the forests, where they may get stuck if the ice on top of the snow is not strong enough to bear their weight. Summer weather that is too hot could dry up watercourses and grazing areas, driving reindeer towards cooler areas with patches of snow or towards the shade provided by forests, which generally limits their grazing possibilities. External disturbances such as the logging of forest areas limit reindeer's ability to adapt to warmer summers as well as their access to arboreal grazing in winter. On the other hand, the recent fall in the price for reindeer meat showed reindeer herders that good years are not unproblematic, as they had no system or capacity at their disposal to handle excess production from the good years (cf. Stakeholder meeting, reindeer herding, Norway).

Reindeer herders generally speculated that adaptation is possible although constrained: 'Reindeer herding will probably continue as long as people want to do it, but you can get extensive restructuring and you can get large fluctuations in profitability and in how many people can make a living from the industry' (Reindeer Herder II, Norway).[164] Climate changes would affect the underlying operation of and interact with political and economic factors. One actor summarized the situation thus: 'If forestry logs a lot but there is a favourable climate and so on, that can sort of compensate. But if climate and forestry and these additional disturbances also work against us, that makes it considerably more difficult' (Reindeer Herder III, Sweden).[165]

Yet adaptation is not only dependent on natural conditions, but something that can be changed depending, for instance, on whether a herder introduces supplementary feeding (moving towards farming), which in turn depends on planning and management. Just how complex the interplay of weather and economic factors is can be clearly seen in the fact that the three years of favourable weather resulted in the production of more meat than the market could absorb, lowering prices (Reindeer Herders' Association, Finland). Measures exist to deal with existing weather trends and with economic and internal political changes that manifest themselves economically, but one can see that economic adaptations are likely to cause reindeer herding to move more towards farming and thus to detach itself further from the natural environment. The natural feeding grounds are compromised by forestry and other land uses, limited in size, and costly to herd and gather reindeer on. Weather conditions may also limit the availability of grazing areas, decreasing herders' income further.

4.6. The vulnerability of reindeer herding

Vulnerability in the reindeer herding sector is the result of multiple factors, most of which lie in the international market situation, national regulation (and possibilities for enforcement) and climate/weather conditions. Supplementary feeding as a management and adaptation system mediates many of these effects and constitutes a mechanism allowing individual reindeer herders to increase control over production and their livelihood. To a large extent, however, reindeer herding suffers from a poorly developed market that diminishes the value of investments aimed at making production more efficient. The sector is also impacted by limited recruitment into the livelihood, largely a result of declining income. Many of the interviewees also worried about the situation in the area at large in the future, given current trends of a declining and aging population and outmigration.

The most positive attitudes were found in Norway, where the reindeer herding economy is strongest. Adaptation will be possible if changes are not too rapid or do not make reindeer herding too expensive, as the additional costs associated with climate change may do (Reindeer Herder III, Sweden). Interviewees emphasized, however, that there is a limit to adaptation: 'In a certain situation you can see that the industry cannot adapt anymore, but right now we do not know where that limit is, whether we are close to it' (Reindeer Herder III, Sweden).[166] The ultimate 'option', if the situation becomes too costly for the individual actor, is 'elimination' (Reindeer Policy, Sweden),[167] where herders are 'out of the contest' and forced to take up employment in different economic sectors or geographical areas.

5

Perceptions of Change, Vulnerability and Adaptive Capacity among Fishing Stakeholders in Northernmost Norway

5.1. Introduction: Fishing organizations in Norway

Historically, fishing has been one of the basic economic activities along the Norwegian coast (Seppänen, 1995). The Norwegian fishing industry, including aquaculture, is one of the largest export industries in the country, accounting for 5–10 per cent of total exports when fish farming is included (Selnes, 1995; Aasjord, 2002). Both the ground-fish and pelagic fisheries are traditionally important. The principal species economically for sea fishing is cod, but haddock, capelin and herring are also important (Hoel, 1994).

The fish resources in the Barents Sea have varied considerably in recent decades, often due to a combination of overfishing and environmental conditions that affect the fish food chain and conditions for fish. Among the relevant conditions is the location of the polar front, which is where the more salty and warmer water of the Atlantic Ocean meets the less saline, colder water of the Arctic Ocean to form a zone that is conducive to production (Gjøsaeter, 1995; Lange, 2001). Examples of the variability in fishing include the depletion of the herring stock to near extinction during the 1960s, after which much effort was focused on capelin. Capelin supported large fisheries in the 1970s, but declined to a very low level in the mid-1980s and again in 1992–1993 following a recovery (Gjøsaeter, 1995).

When fish populations fluctuate due to natural causes and overfishing, the communities along the coast that are dependent on a stable income are directly affected (Gjøsaeter, 1995; Baerenholdt, 1996). 'Northern Norway has half of the total population dependent on fisheries in Norway, while the number of north Norwegians is only about one tenth of all Norwegians' (Baerenholdt, 1996, p227). In 1994, about 10,000 full-time fishermen (mainly men) and about 5000 people employed onshore were active in the fishing industry in north Norway; in other words those involved in fishing, fish processing and trade in the related equipment, and related practices such as fishing vessel repair, services and the public sector ultimately dependent on fisheries represented about half of the local workforce (Hoel, 1994). In 2002, these numbers had declined to about 6700 full-time fishermen, with proportionate – and sometimes larger – declines in processing (Aasjord, 2002), but there remained a large dependence on fishing.

In Finnmark, the northernmost county in Norway and the area focused on here, the production value of the fisheries sector is significant. In 2000, it had a 'first-hand' (i.e. pre-wholesale) value of approximately NOK 2600 million. During the last 15 years, the fishing fleet has been reduced by almost half to the point where it now comprises some 1700 vessels and about 1500 full-time fishermen (Finnmark County Administration, 2000). The catch in Finnmark accounts for 10 per cent of national fish production; however, about 50 per cent of the fish deliveries to Finnmark are now made by Russian trawlers (Finnmark County Administration, 2000). Fishing in northern Norway also has certain characteristics that give it an impact on a large number of people. In contrast to fishing further south in the country, fishing in northern Norway is relatively small-scale (Selnes, 1995). The fisheries sector employs about 10 per cent of the workforce in the county (Vilhjálmsson and Hoel, 2005). Fishing in Finnmark uses small vessels to a large extent, many of which are smaller than 15 metres in length (Finnmarks Fiskarlag). In northern Norway, fishing is also traditionally important for local groups including the Saami. A considerable Saami community defines itself as the Sea Saami, who are traditionally dependent on fishing rather than on reindeer herding and are a very important segment of the population in Finnmark County. As a result, the Sametinget, a Saami-elected advisory body to the state, is also highly involved in fishing.

Administratively, the fishing industry is regulated on many levels, predominantly the international and national. Regulation has been largely developed to limit the allowable catch or quota for different levels and actors in order to maintain sustainable levels for the shared catch in the Barents Sea. About 80 per cent of the Norwegian catch originates in fisheries that are shared with other countries. To preserve the resource basis, fishing levels are set internationally by the International Council for Exploration of the Seas (ICES). Based on these catch levels, Norway negotiates annually with different parties about catches and fishing rights and access to fishing in Norway's economic zone. The most important negotiations are between Norway and the EU; Norway and Russia (in the Norwegian–Russian Fisheries Commission); and Norway, Iceland, the Faroe Islands and Greenland. At the national level, decisions on regulations are made by the Ministry of Fisheries and Coastal Affairs (Fiskeri- og kystdepartementet) and the Directorate of Fisheries (Fiskeridirektoratet). The ministry has the main responsibility for managing and controlling the catch, while the Directorate of Fisheries – the ministry's main advisory and executive organ – has various responsibilities related to resource management, regulation and control, and maintains a number of regional offices (cf. Hallenstvedt, 1993). At the time of the study, regulations for determining quotas for different groups of fishing vessels had been developed by the Reguleringsrådet, a consultative body to the

Directorate of Fisheries. This body included representation from sea research and fishing authorities, the fishing industry, environmental agencies, and Saami fishing interests; today, the Reguleringsrådet has been replaced by ongoing meetings between the Directorate of Fisheries and the different interests. Regulation takes place through different types of quotas that were instituted in 1990, such as total quotas, group quotas, maximum quotas and vessel quotas (Hallenstvedt, 1993).

Typical of the organization of fisheries in Norway is the existence of an organization controlling the first-hand sales of fish. The Norwegian Raw Fish Organization (Norges Råfisklag) is the dominant sales organization; it negotiates and determines the base prices and may also subsidize the transport of fish to buyers (Krogh, 1995). There also exist smaller sales organizations, such as the Norwegian Fishermen's Sales Organization for Pelagic Fish (Norges Sildesalgslag). Moreover, the Norwegian fishing industry is influenced by a number of interest organizations, of which the dominant one is the Norwegian Fishermen's Association (Norges Fiskarlag). It is organized into a number of county-based districts, such as Finnmarks Fiskarlag for Finnmark County. The national organization for coastal fishers – a less influential organization than Norges Fiskarlag – is called Norges Kystfiskarlag.

This chapter describes, among other things, this network of fisheries organizations as it is perceived from a local point of view among interviewees in small-scale fishing in Finnmark County. The material illustrates that fishing is an intensively regulated sector, with the distribution of quotas a major discussion point. The quota system has made the primary resource – even more so than in forestry and reindeer herding – into a strongly regulated area rather than the commons it once was, with impacts on a fishing industry that developed well before the introduction of the system and on the resource access of different groups. While the rationale behind a quota system is not disputed among interviewees, they challenge the way in which quotas have been distributed among small- as opposed to large-scale actors; local small-boat fishermen and local organizations attempt, for instance, to influence the decision-making on quotas and regulations through international norms that have implications for national regulation.

The chapter discusses, first, the stakeholders' perceived socioeconomic changes in fishing, which are centred on the quota system and its economic implications for fishermen. International market changes and competition from countries with lower labour costs also exert a notable influence on fishing. The chapter then outlines the political changes and governance network of fishing, which in the fishermen's opinion entails a strong conflict between small- and large-scale fishermen and their respective organizations. The description encompasses the local, regional and national levels of decision-making, national interest groups, and the impact of international norms and the international and EU levels. Third, the chapter describes the possible impacts of climate change and the context of environmental changes that actors noted for the area. Perceived environmental changes that affect fishing include the large growth in king crab, sea urchin and seal populations, which impact coastal underwater flora and fauna and thereby the conditions for fishing. The concluding sections of the chapter then describe the adaptations regarded as viable for dealing with the threats described. To a large extent, these adaptations include higher-level and regulative measures for dealing with the distribution of quotas and the control of, for example, the king crab population. While the chapter primarily addresses coastal fishing, the interviewees include river salmon fishermen who work at, among other places, the mouth of the River Tana. These examples provide a wider context illustrating socioeconomic change in the area and vulnerability in small-scale traditional activities. Given that this chapter builds on only a single case study, the results are merely indicative of the specific vulnerabilities and limits to adaptive capacity that may exist for the targeted sector and area.

5.2. The socioeconomic change in fishing

Fishing in northern Norway has seen considerable changes, particularly since the introduction of the quota system in 1990. The relevant fish-processing system for the market has changed from fillet production to fresh fish production, resulting in a major restructuring of the northern Norwegian economy. Northern Norway, traditionally characterized by relatively small-scale fishing and combination subsistence, as discussed below, has changed during the last generation towards more large-scale economic activity and a rationalization of fishing.

5.2.1. Change over time

Combination subsistence has been relatively common for a long time in the areas studied. From a basis in fishing or small-scale agriculture, people have practised fishing, hunting, berry picking and household logging (Coast Fisher I, Norway; Coast Fisher II, Norway): 'Not all had boats for winter fishing … It was necessary to do something else on the side … then it was the land that was relevant' (Coast Fisher II, Norway).[1] At the mouth of the Tana, seals were hunted for food and for fur (Coast Fisher I, Norway). Many engaged in inland or river fishing in the 1960s and 1970s to supplement their income from sea fishing (Coast Fisher II, Norway). Salmon fishing was practised in the River Tana (Coast Fisher I, Norway).

Although some people still practise combination subsistence, the number engaged both in this and in fishing generally has fallen (Coast Fisher I, Norway). The number of active salmon fishermen in the case study area declined from some 40 to 50 in the 1970s to some 10 today (Coast Fisher I, Norway). In the case of sea fishing, many smaller boats have been taken out of operation. Interviewees noted the relatively short time within which the focus has changed: 'You catch the same amount of fish today [in Norway] with 15,000 fishermen as you did with 60,000 … some ten years ago … The pace of development, technological development, has been furious' (Finnmarks Fiskarlag).[2] 'These primary industries have disappeared very quickly in these northernmost municipalities such as Tana and probably others … there has been such unbelievable urbanization' (Coast Fisher III, Norway).[3] Owing to better communications, some fishermen have also moved inland (Local Government I, Norway). The expenses associated with fishing, for instance the cost of fuel, have also increased (Local Government I, Norway).

The fishing industry and its character have changed correspondingly. In Finnmark, the fisheries expanded vigorously in the post-war era through the 1950s and 1960s, peaking at the end of the 1970s, through the development of boats, equipment, and the fillet and freezing industry. This development transformed fishing into a year-round industry (rather than one relying on a coastal fleet that caught its catch mainly in spring). At the time many of the interviewees started working (in the 1960s), fish was mostly refined into dry fish by being hung to dry on large racks or was put on ice and sent south with the regular Hurtigruten ferries travelling from Finnmark to Bergen (Coast Fisher II, Norway). Over time, further processing was developed through freezing installations, where fish was frozen for processing into fish fillets later in the year. Fillet production quickly became the occupation that provided most of the work in the fisheries sector. However, this focus on large-scale production also precipitated the structural change seen in the fishing fleet starting in the 1960s and 1970s. Finnmark County traditionally had many small boats, in which fishermen sometimes worked alone and would often come back to the coast daily (Coast Fisher II, Norway): 'In Finnmark there is more of a tradition of fjord fishing. It has not been capital intensive and everybody in the area has had the right to use the maritime resources' (County Council, Norway).[4] As the fisheries industry grew, however, 'the fishing fleet was not able to deliver enough raw material to run it year-round, so trawlers started to be built' (Directorate of Fisheries, Norway).[5] This situation contributed to some extent to a state policy favouring larger-scale fishing to the point where today 'boats have been decommissioned … structural policy has been applied to decrease the number of boats and companies' (Directorate of Fisheries, Norway).[6]

At the same time, the restructuring in fisheries since the late 1970s has also been impacted by changes in catches. A coastal fisherman noted: 'You got a relatively good price for both flounder and halibut … but they also became fished out very quickly when they started fishing with nets and very large boats stationed in the fjords' (Coast Fisher III, Norway).[7] Capelin populations, which were comparatively important to the areas, dipped at the end of the 1970s, as did cod populations in the 1980s, with the latter then collapsing in the late 1980s (Coast Fisher III, Norway; Fishing Interest Organization, Norway).

To some extent as a result of these fluctuations and in some instances overfishing, quotas (based on the so-called individual quota system) were instituted for cod for conventional vessels starting from 1990. In order to maintain fish populations, the quota system limited access to fishing as an occupation in practice. Interviewees expressed sharp criticism of the regulations through which this has taken place. In order for a fisherman to gain a vessel quota after 1990, he or she had to have caught a certain amount of cod in the previous three years (relative to boat size, among other things). This resulted in difficult situations in Finnmark, where Eastern Finnmark had had seal invasions in the ten years before the introduction of quotas that had limited catch size during those years (Coast Fisher I, Norway; Fishing Interest Organization, Norway). Interviewees perceived that many people lost their fishing rights because of their limited catches in the years preceding the quota system, and emphasized their loss of rights in this instance and generally. As one actor observed: 'The largest change as I see it is the rights. They are disappearing. The biggest mistake … that was the quota system in cod fishing … for the fjords' (Coast Fisher II, Norway).[8]

Moreover, in the 1990s Finnmark had an explosive increase in the king (Kamchatka) crab population, which moved into the area from Russia. Today the crab is both an important species and a problem:

> *[The king crab is] a pest and a pain and a blessing. For those who have a quota it provides a very good income during a month in the autumn when there would otherwise be little else … But for those who use nets in spring to fish for cod or lumpfish, it is devastating … They get enormous numbers of … ruined nets.* (County Council, Norway)[9]

Among the requirements for obtaining a fishing quota for king crab, however, is that the fisherman has a quota for fish and a boat of a specified minimum length. For this reason in particular, some interviewees regarded the king crab as contributing to competition within fishing: 'The king crab is now so numerous in the fjord that you cannot use nets for cod fishing anymore and you cannot fish for lumpfish with nets; it is impossible to meet your quota and people are left out' (Fishing Interest Organization, Norway; cf. Coast Fisher III, Norway).[10]

As a result of these diverse changes, fishing in Finnmark has undergone extensive restructuring. In the early 1990s, 'in addition to the resources being depleted there was a large wave of bankruptcies here in Finnmark' (Directorate of Fisheries, Norway).[11] The increase in Russian catches sold to Norway following the fall of the Soviet Union 'gave the fisheries industry new legs to stand on … secured raw material. But in 1995–1996 there was another wave of bankruptcies' (Directorate of Fisheries, Norway).[12] Fishing at the time also started to suffer from a limited competitiveness with emerging markets in the area of fish processing, such as those in Russia and China:[13]

> *What happened in the 1990s on the fisheries side was that you have gained an entirely different turnover of fish. Firstly, it was earlier prohibited to freeze fish on the boat but now there is actually permission … to do so. During the 1990s … 'freezing hotels' were established … freezing stockrooms for fish. You can freeze the fish on board and deliver*

> *it to freezing stockrooms and deliver it on to those who bid the most money, partly according to auction principles.* (Directorate of Fisheries, Norway)[14]

As frozen fish and fillet production became less dependent on locality than previously – in other words fish could be processed far away from where they were caught – stronger competition for fish resources arose (Directorate of Fisheries, Norway). As a result, fish reception points, freezing installations and the fillet industry did not receive as many deliveries as before, although they were somewhat helped by the beginning of deliveries to Finnmark from Russian trawlers (which nevertheless ultimately competed with Norwegian catches) (Directorate of Fisheries, Norway):

> *It was during those years that the market became very much the focus, because at that time you realized that it was not the quantity that was the problem; it was the prices – pure and simple; and perhaps the fillet industry, which is so extensive, was not really as competitive anymore.* (Directorate of Fisheries, Norway)[15]

Fisheries in Finnmark have thus witnessed cyclical and rather large-scale change as a result of regulation as well as variability in and accessibility of fish stocks and market fluctuations. One interviewee, however, noted that the reason for the decline in fisheries production might have been that the expectations of the fishing industry in northern Norway regarding employment and large-scale local refinement have been too high:

> *This focus on investment in companies that operate year-round and [the notion] that fishing should be a year-round industry like any other perhaps had too large ambitions given how much it [the situation] varies. The fisheries have not been able to sufficiently adapt to a varying framework.* (Directorate of Fisheries, Norway)[16]

During the last few years, continual adaptations to the market and the changing resources have often taken the form of the least profitable actors being driven out of business: 'We [had] a wave of bankruptcies last year again ... some companies never came back very well after 1995–1996. The last three years there has been a rather marked decrease' (Directorate of Fisheries, Norway).[17] As a result, some fishermen have started diversifying, for instance through tourism: 'There has been established winter tourism here based on snowmobile rides and trips to the fjord to catch your own king crab ... There are no other places in the world where this can be offered' (Directorate of Fisheries, Norway).[18]

Fishing today is thus marked by larger, more effective boats (Directorate of Fisheries, Norway), a short but intensive fishing season (utilizing these more effective means) lasting until the quota is caught (County Council, Norway), and discussions of the rights to resources in the sector following the introduction of the quota system in 1990, which interviewees predominantly regarded as benefiting large-scale rather than small-scale fishermen. Fishing entrepreneurs in the Finnmark area have, over time, decreased in number and their businesses have sometimes become larger, while some have left the business or diversified (Coast Fisher II, Norway). At the same time, the fishing industry has changed in character into an increasingly international business, where prospects for fish processing are strongly influenced by the international market and production possibilities elsewhere, increasing local vulnerability to large-scale economic and political changes and requiring fishermen to adapt extensively to the market.

5.2.2. The quota system and fishing economy

In today's quota system, introduced in 1990, quotas – the right to fish – are assets that are bought and sold rather than open access rights (Coast Fisher II, Norway). Quota sizes differ depending on fishermen's previous catch and boat size, among other things. One fisherman pointed out that 'there are large differences [between category 1 and 2] ... I have a boat in the 8–9-metre category and I have a cod quota of 17–18 tons ... and then I have a quota for haddock and saithe in addition. Those who have the same boat length but a category 2 [quota] have a catch of some 7–8 tons of cod, so it is less by half' (Coast Fisher II, Norway).[19] To provide some flexibility in the potential catch for smaller fishermen, depending on what is available, the quota is expressed as a combination quota (*samlekvote*) for cod, haddock and saithe, so that fishers can vary their catch depending on price and availability. For example, 'if the price and availability of haddock is better than that of saithe, you can take somewhat more haddock and less saithe' (Directorate of Fisheries, Norway).[20] Fishermen may also obtain a king crab quota (Coast Fisher II, Norway).

For the individual fisherman, the economic problems lie in the difficulty of obtaining a quota and, especially for young fishermen, of starting out in the business. Those who already possess quota rights – especially category 1 rights – are in a relatively favourable economic situation, which illustrates the differential vulnerability among small-scale fishermen. One interviewee pointed out that a combination quota of 40–50 tons for cod, haddock and saithe would make it possible for someone to buy a boat and a quota for a million NOK:

> *That quota would be sufficient for you to justify the investment ... But then you have equipment, so if the equipment does not come with the boat, you have to put up the money and it quickly rises to a couple of hundred thousand NOK. So it is a prerequisite that you get 50 per cent support to start fishing, but then it is worth the effort ... if you also get a crab quota, you will have a good surplus. But someone who is just*

> *starting out has not earned the rights for a crab quota.* (Coast Fisher II, Norway)[21]

As an example, a small boat under 8 metres may have a quota of 16–17 tonnes, which allows the owner to earn some NOK 200,000–250,000 before tax (Coast Fisher I, Norway). One interviewee noted that he could earn as much with a crab quota alone and that once a person is established as a fisherman and has both a fish and crab quota, he or she has no financial problems (Coast Fisher II, Norway). However, one interviewee noted that 'the only chance to get a category 1 quota for cod fishing is to buy a cod quota, but that is not profitable. You have to work for free for ten years, or, in other words, you work for the bank' (Coast Fisher I, Norway).[22]

As a result of rationalization and the problems of starting out as a fisherman, the recruitment to fisheries in Finnmark has decreased. 'Earlier up to four or five [people] worked on each boat; today there are rarely even two. There is normally one fisherman per boat. The rationalization of fishing has meant that fewer people can make their living from it' (County Council, Norway).[23] Through rationalization, employment is replaced with technology: 'What we do is capitalize this. We are exchanging people for money all the time. We put in money, and buy ourselves technology, and have to give people notice because we cannot afford them' (Norges Kystfiskarlag).[24] This has impacts on the local economy as well as on the sector at large, and there are some concerns that the knowledge of fishing is declining and that the population is decreasing in Finnmark: 'The competence disappears and it will be older people [who are active in fishing] and when they quit they sell their quota as a form of pension insurance' (Saami Parliament, Fishing, Norway; cf. Coast Fisher I, Norway; Coast Fisher II, Norway).[25]

Some interviewees also noted the impact of financing institutions on fishing as a factor complicating recruitment and employment. One actor noted that, in general, obtaining bank loans for small-scale fishing investment was 'close to hopeless' (Norges Kystfiskarlag).[26] Partly, this may be because of the capital required of the borrowers in order to obtain a loan; for example, they have to put together perhaps half of the cost of a small boat on their own. Partly it may be on account of the overcapacity in fishing: '[there is] overinvestment. There have been investments in trawlers and in purse seine vessels, billions per year. It is the fish that have to pay for all of this. There have been investments of some NOK 8–9 billion in recent years' (Norges Kystfiskarlag).[27] Others also noted that the banks place considerable demands on short-term profitability, which may make them require accelerated repayment during a market slump in order to receive returns on their investment, causing bankruptcies (Directorate of Fisheries, Norway): 'The banks are very short-term; they get so bloody afraid if things start going wrong early, but we know that this industry varies. They have too little patience' (Finnmarks Fiskarlag).[28]

5.2.3. Market adaptations: Different types of fish processing

In the current situation, where fish quantities are limited by quota allocation, many efforts are made to add value through new types of processing, in other words to gain a higher income per unit of catch and increase profit possibilities. Traditional processing of fish by drying is still practised in the areas: 'The prices of stockfish have been good, increasing since there are fewer who dry fish' (Coast Fisher I, Norway).[29] The hygiene and facilities required for producing dried fish have been improved, raising quality but also increasing production costs; at the same time, however, higher prices can be obtained for the higher quality dried fish (Saami Parliament, Fishing, Norway). King crab has become an important cash cow for many fishermen, as it is both sold in Norway and extensively exported to Europe and beyond (to Japan, among other places, where there is a large market) (County Council, Norway). In contrast, fillet production, which for a time was the dominant processing method, has declined due to competition from countries with lower labour costs yet advanced technology (Saami Parliament, Fishing, Norway). As a result of these changes in distribution and products, fish processing in the Tana area is very limited today (Local Government I, Norway). The processing plants that exist are relatively small-scale: 'They have been the most flexible and have had lower fixed costs, investment costs' (Directorate of Fisheries, Norway).[30]

One response to this situation has been the development of fish farming, and here some people have great hopes for the future. As an operation that controls resource production rather than relying on a catch, fish farming can avoid the problem of large natural fluctuations. It may provide support in particular for smaller local communities in the area, even offering significant employment opportunities (County Council, Norway): 'The major [emphasis] is salmon farming. It is important [already today] … for the local areas that have it' (County Council, Norway).[31] Thus, in fish farming, 'throughout the 1990s … production actually rose tenfold in a few years and now we are at some 50,000 tons [here in Finnmark]' (Directorate of Fisheries, Norway).[32] This development has largely been beneficial for the area, with the exception of wild salmon fishing, which has decreased markedly as cheaper, farmed salmon has driven down the prices.[33] Fish farming, however, has also had its downturns, in particular a wave of bankruptcies in the late 1990s. In the opinion of one interviewee, this was the result of strict demands for loan repayment by financing institutions rather than of the market situation (Directorate of Fisheries, Norway). Interviewees also suggested that in the future both farmed fish and regular catches could increasingly be exported to Russia, which they saw as emerging from its economic crisis as a major new market where Norway may have geographical advantages (even if a somewhat high price level) (County Council, Norway).

In a similar vein, the most recent major adaptation to the market has been the development of fresh fish sales as a means to compete through quality rather than quantity. This development can be attributed to both the decreasing access to resources caused by the quota system and the increased competition internationally where price and processing are concerned. One actor expressed this very clearly:

> *There are two factors that impact my income. These are quantity and price. If I catch, say, 35 tons of fish instead of, say, 130 tons in one season … it is clear that we have to have a better price. That's why the fillet industry is a low-price industry all the time. It is the amount that should carry it, and, given the level of costs that we have in Norway today, it is quite simply not profitable … with the low quota we have.* (Norges Kystfiskarlag)[34]

Fresh fish is thus increasing as an option for creating added value; its market benefit lies in its freshness and in just-in-time delivery, which provides for a relatively high price level and a product where fewer actors today are able to compete (Saami Parliament, Fishing, Norway). Production of fresh fish is suitable for small-scale fishermen, as small boats regularly come in to port every evening and can therefore empty their live catch into aquaculture units, where fish are kept alive and fed until they are delivered to buyers (Saami Parliament, Fishing, Norway). Some consider this handling of fresh fish as demanding enough for it to be regarded as a form of processing (Saami Parliament, Fishing, Norway) that constitutes a middle ground between fishing and fish farming and partly avoids the relative unpredictability of fishing and the risk that catch times and levels are not attuned to the market: 'You can guarantee the delivery time, you can guarantee quality … You can feed [the fish] and then you can keep them for a year' (Saami Parliament, Fishing, Norway).[35]

As a system under development, fresh fish production is also to some extent an adaptation to the problem in Finnmark of limited fish reception points, that is places where buyers pick up the fish for further sale (Coast Fisher II, Norway; Local Government I, Norway). Fish reception points are often far from one another, and pick-up times differ at those that exist (Saami Parliament, Fishing, Norway). The benefit with fresh fish is that it is less dependent on pick-up times (Saami Parliament, Fishing, Norway; cf. Directorate of Fisheries, Norway; Coast Fisher I, Norway). Although there are plans for increasing the number of fish reception points (Coast Fisher II, Norway), the feeling is that the dependency on reception points may decrease with increased fresh fish production (Saami Parliament, Fishing, Norway). New communications technology also plays a role in the process of selling fresh fish, and may further decrease the dependency on reception points and sales agents: 'The internet has resulted in a bond of sorts between fishermen and the market; they have become mindful of each other. They have gained the capacity to communicate in an entirely different way' (Saami Parliament, Fishing, Norway).[36]

This may result both in more appropriate, finer-tuned market adaptation and greater perceived responsibilities and ties between buyers and individual fishermen.

As buyers know the origin and quality of the fish, individual fishermen are able to sell directly to them, receiving a higher price and thereby a higher income from their limited quota (Saami Parliament, Fishing, Norway):

> *There is a lot of money involved in the new way of adaptation ... There are private organizations that want to come in and trade ... and when a fisherman who is on the way up in the business, who has the requisite knowledge about this and ... has a quota, he wants the most he can get out of his quota ... [Maybe] he catches his quota's worth and has it penned and uses his leisure time or other time off to sell it.* (Saami Parliament, Fishing, Norway)[37]

On balance, interviewees described how fishing has changed over time to become increasingly large-scale and regulated. The clearest and most frequently discussed example of regulation was the quota system. The market and market mechanisms, on the other hand, were largely taken for granted: adaptation to changing market requirements, such as those resulting in the closure of fish-processing units and their being moved to lower-cost locations, were mainly discussed in terms of needs for adaptation, not in terms of action or protest (unlike the reaction to quota setting). Accordingly, different types of fish 'processing', through fresh fish production, were discussed as ongoing adaptations. This situation illustrates the relatively extensive integration of fisheries in the global market: fishermen must compete with fish production and price levels worldwide, resulting in a relatively high but differential local vulnerability to changes that take place beyond the locality and even the country.

5.3. Political and organizational change in fishing

This situation of ongoing change in the fishing sector is reflected in changes and conflicts at the political and organizational level, primarily regarding the quota distribution system. Here, local and regional government and organizations were seen as conflicting with national interests. The national interests were seen as having the largest impact on policy, although local and regional levels are significant in supporting small-scale fishing and developing a relatively united regional interest in Finnmark.

5.3.1. Contested fishing rights

The control of fishing rights through a sellable quota is by far the issue that was most intensively discussed by interviewees. Their observations centred on the right to catch fish, how this right should be distributed, and how it can be guaranteed that the region and the people who live there will retain the legal and moral right to fish. For instance, the Saami Parliament noted that 'the main problem is the rights, because if there are no rights then it is but a fairytale, all of this' (Saami Parliament, Fishing, Norway).[38] Other interviewees in Finnmark put things in equally strong terms: 'If [today's policy] continues ... for 10 more years, there will be no fishing here in 20 years. Not one person [in the business]' (Coast Fisher III, Norway).[39] A small fishing interest organization described the situation as follows:

> *People in this area have practised fishing since time immemorial and Finnmark is the area of the country that has the oldest recorded population. It is the oldest registered [settlement] according to carbon 14 dating, 10,300 years on Magerøya ... If we take 10,300 years ago as the point of departure, people have had the right to use resources for 10,285 years, whereas this right has been restricted for the last 15 years* (Fishing Interest Organization, Norway).[40]

Similarly, a fisherman noted that 'earlier, you called it an "open commons" and that meant that everyone had the right to fish. Now it is closed, but what happens is that we now see control by owners within the fisheries, which only seems to be getting stronger. That the big should get bigger. It is close to being a political aim' (Coast Fisher III, Norway).[41]

However, none of the interviewees directed their criticism primarily or explicitly at the quota system as such.[42] Rather, they criticized the policy on the distribution of quotas to different actors (Saami Parliament, Fishing, Norway; Coast Fisher II, Norway; Coast Fisher I, Norway). The main concerns were the 'devolution' of quotas – and thus the rights – from Finnmark, with its preponderance of small vessels, to other regions in Norway, through what was seen as a national policy favouring larger-scale fishing. This was implemented in the so-called *kondemneringsordningen* (decommissioning order) that phased out boats under 15 metres in length (Coast Fisher II, Norway; Saami Parliament, Fishing, Norway). The decommissioning order operates such that a fisherman who sells his or her quota back to the state receives tax-free compensation for the quota, which amounts to almost a year's salary in some cases. The boat is destroyed or sold as a recreational vessel (with somewhat lower compensation paid to the fisherman), and the quota goes back to the state for redistribution (Coast Fisher II, Norway).

This order was widely discussed by interviewees as decreasing the small-boat quota in practice. Interviewees in general criticized the state policy as one that would decrease the population in northern Norway: 'In the extreme case some 20 vessels will be able to catch the entire Norwegian cod quota, you might say. There is no need for people to live here ... but where is there a need for people to live? Should we all huddle together [in larger cities]?' (Norges Kystfiskarlag; cf. Norwegian Raw Fish Organisation).[43] Many actors also emphasized that this situation, in which larger boats catch more fish, also affects where fish are brought to port and thus employment on land. Large-scale fishing often brings its catches to ports further south, with negative impacts on regional employment and infrastructure (County Council, Norway). One interviewee noted that the owners of trawlers fishing off the coast of Finnmark are 'mainly companies from Vestland [region]. They take the fish down there because they get a somewhat better price… the industries here in Finnmark are caught with their pants down. This is why everything is going bankrupt … here in Finnmark' (Coast Fisher I, Norway).[44] Similarly, one actor stated:

> *This is the way it is everywhere, if you look at food market chains, they should be bigger and bigger, they are merged, everything should just be bigger and bigger. It must be big enough, then it is profitable. Earlier it was that … you should [use fishing as] … a means of maintaining [population], in a way that no one speaks about anymore. We have terms such as the proximity principle … this is no longer discussed.* (Norges Kystfiskarlag)[45]

Additionally, as quotas can be bought by fishermen in other regions, there is a concern that fishing rights may disappear from Finnmark County, even with regulation in place to limit the devolution of quotas from the north (Directorate of Fisheries, Norway). For instance, a coast fisherman noted 'that is the way it is everywhere – if you have money enough you will get [the quota] as you wish. That has never been a problem' (Coast Fisher III, Norway).[46] There was also a worry among some of the interviewees that in the longer run quotas may be owned not only by southern Norwegian interests but by foreign interests, particularly if Norway should join the EU (Coast Fisher III, Norway). Such larger-scale actors could have an even more sweeping effect on local fishing in Finnmark: 'It is a given that the larger units we get, the more dramatic all future actions will be for the areas' (Coast Fisher III, Norway).[47]

Consequently, some noted that the conflict over fishing rights, clearly articulated in interviews, was prompting a united front and:

> *… a political shift of gears … perhaps foremost in some of the parties here in Finnmark, it seems that they have noticed that the fishing policy has been wrongly targeted in relation to the coastal towns … the population is declining to the extent that in 20–30 years there will be only a few locations in Finnmark with people living in them. It is entirely clear for county politicians now that … this not a desirable development.* (Fishing Interest Organization, Norway)[48]

5.3.2. Local and regional-level support for fishing

As suggested above, local and regional actors are relatively united against what is seen as a national policy disadvantaging local, small-scale fishing in Finnmark. For instance, the Saami Parliament strongly emphasized that fishing rights should remain in the area. The Saami Parliament does not distinguish between fishing rights for the coastal Saami (known as the Sea Saami) and the rest of the local population, but argues for fishing rights in the region as a whole: 'We have fought for those who live here to have the right to fish' (Saami Parliament, Fishing, Norway).[49] 'All of those who live in the area have equal rights ... when it comes to defending these rights, you may say that it can be based on Saami history, but when it is a question of [rights] in the area, these are shared equally' (Saami Parliament, Fishing, Norway).[50] Actors also noted that the distinction between indigenous and local is often difficult to make, as large segments of the population have long traditions in the area and are often of unclear or mixed descent (Fishing Interest Organization, Norway). While the justification for fishing rights is seen as based in indigenous rights, it is ultimately applied to the area and population as a whole, depending on their involvement in small-scale fishing.

Many of the actors further noted political aims to change the quota distribution and thereby improve the recruitment situation in fishing. For instance, Finnmark County Council noted that it would 'attempt to promote the interests of the smaller boats that land fish locally and pay their tax to the municipalities in the area and keep the areas populated. Smaller boats mean that there are more people who can make their living from the fisheries industry' (County Council, Norway).[51] The County Council discussed, for instance, the development of collective quotas:

> *One way of retaining the rights in the area is to start a resource company that buys a quota and rents it to fishermen. Regional quotas which are managed by the county or a similar organization would keep the rights to fishing in the county.* (County Council, Norway)[52]

Similarly, the Saami Parliament interviewee stated: 'We imagine that we will have regional resource management where the region has its quota ... and based on that can manage recruitment' (Saami Parliament, Fishing, Norway).[53] One of the results of such ambitions is the local project Tanafjordprosjektet, which includes both local and regional policy efforts and places 'considerable importance on a resource rights protection law for fishing, because that is something that is rather deplorable today' (Fishing Interest Organization, Norway).[54]

The local and regional levels are also able to offer some support to local small-scale fishing. Municipalities have a development fund for fishing, and fishermen can also apply for municipal funding when buying boats and fishing equipment (Local Government I, Norway). At the regional level, the county provides resources for fishing support, although at a decreased level: 'Earlier extraordinary support measures had a larger role and the county level … had more support measures for those kinds of things … now they no longer have that' (Directorate of Fisheries, Norway).[55] Finnmark Fylkeskommune supports fisheries' investment through an investment fund and can also support recruitment in fisheries by financing 10 per cent of the cost of a boat (County Council, Norway). The Saami Parliament supports small-scale fishing recruitment by providing a development fund offering support of 20 per cent towards the cost of a boat within participating municipalities in the Saami management area, up to NOK 100,000 for old boats and NOK 250,000 for a new boat (Saami Parliament, Fishing, Norway). The Saami Parliament further supports combination subsistence through a yearly NOK 25,000 subsidy for each practitioner found eligible (Coast Fisher I, Norway). Using all these available funds, a fisherman starting out may obtain some 50 per cent of the cost of a small used boat with a quota (Coast Fisher II, Norway). Municipalities also attempt to enhance employment, for instance through favourable harbour facilities that attract fishers to bring in their catches there (Finnmarks Fiskarlag).

The local and regional lobbying efforts have had some success in influencing policy and, especially in discussions of special rights for Finnmark, a developing body of law known as *Finnmarksloven* (established in 2005). Through these discussions, 'the Department of Fisheries has now reluctantly gone along with us, but how long they will I do not know yet' (Saami Parliament, Fishing, Norway; cf. Coast Fisher I, Norway).[56] The degree to which local and regional structures can influence the political system are very limited, however: 'It is not any local or regional policy that dominates the development in this line of business' (Local Government I, Norway).[57] Similarly, 'The county has never been an actor in fisheries policy' (Saami Parliament, Fishing, Norway).[58] Interviewees thus emphasized that resources at the local and regional levels are limited:

> *It is always national policy that determines people's actions [and] we notice a continuously growing, intensified development where the state has reduced income transfer to the municipalities … it will become more difficult for us to … obtain funding in the future.* (Directorate of Fisheries, Norway)[59]

However, any changes in fishing that limit local tax revenues are likely to further increase municipal economic losses and further restrict local government resources (Local Government I, Norway). There has been a restructuring at the municipal level in Tana, leading to the loss of some 25 jobs during the past few years and decreasing administrative resources in an ongoing process: 'Nationally there are strong forces that want municipalities to merge' (Local Government I, Norway).[60]

5.3.3. National-level state and interest group decision-making and quota allocation

Local and regional actors in general expressed a dissatisfaction with policy at the national level. The criticism mainly focused on national policy, especially the *kondemneringordningen*, on lobbying by financial and large-scale interests and, to a lesser degree, on interest organizations such as the Reguleringsrådet; all were seen as disadvantaging local small-scale fishing. Interviewees also noted, however, that state action and involvement in Finnmark is limited by the special standing of the county. A local government representative noted, for instance, that the possibility of state actions in the county are limited, as 'the state's ownership of land in Finnmark [is] disputed ... the state is careful about using its ownership for a larger degree of privatization of the resources since the rights situation is undecided and the state does not want to provoke anyone' (Local Government I, Norway).[61] The special standing of Finnmark may thus result not only in special regulations and benefits for the county, such as the existing regulation on quota sales from the northernmost counties, but also in concerns regarding state possibilities to act in resource maintenance.

With regard to state policy, state administrators at the regional level largely acknowledged that regulation has not addressed specific regional characteristics. The authorities' role, such as that of the regional offices of the Directorate of Fisheries, is centred on implementation and control, meaning that they have limited possibilities to influence policy: 'What we do is national resource management' (Directorate of Fisheries, Norway).[62] The possibility for actors to have close access to this administration is also decreasing, as some offices have been closed and the number of employees has decreased because of restructuring (Directorate of Fisheries, Norway). Interviewees at the directorate, however, also noted that there are positive examples of state regulation that are not emphasized as much locally as other effects. One is the limitation on large vessel fishing and equipment types (such as Danish seines or purse seine trawls) outside fjords, which has been implemented as a result of reports that coastal cod populations were possibly being reduced by fishing at sea. As a result, 'there have been stringent regulations put on [the larger vessels] in the fjords and restrictions have been established there especially on the use of the Danish seine ... so we hope that that will help' (Directorate of Fisheries, Norway).[63]

Interviewees also criticized national policy, noting that conventional small boats are relatively energy efficient and sustainable and that inefficiency thus should not be the reason that they have not been supported on a policy level (Saami Parliament, Fishing, Norway). Instead, interviewees saw policymaking as influenced by interest groups and financing (Coast Fisher III, Norway): 'The making of policy itself has been badly managed' (Coast Fisher II, Norway; cf. Fishing Interest Organization, Norway).[64] The state policy was generally seen as too limited and too far removed from small-scale fishing concerns:

> *The Parliament is not interested in local politics and works in a little world of its own from pure ignorance … It is the lobbying – it's going on all the time, and the organization of fishing vessel owners is very strong there, and has money and has gained enormous power in this. They get some politicians to believe that if it only becomes large enough, it will be profitable enough.* (Norges Kystfiskarlag; cf. Fishing Interest Organization, Norway; Coast Fisher III, Norway)[65]

Policy was here criticized by interviewees as being made through lobbying and industry power structures. Some criticism was directed here at the Reguleringsrådet, which makes recommendations on the allocation of quotas to different vessel categories, to the effect that it does not adequately represent Saami interests (Saami Parliament, Fishing, Norway; cf. Directorate of Fisheries, Norway). However, alternative models for quota distribution have been discussed, as well as the proposal that when a boat is decommissioned, its quota should remain within the original quota category rather than be redistributed to a pool for all fishing vessel sizes (Saami Parliament, Fishing, Norway).

Some of the criticism of state actions and quota decisions was also targeted at Norges Fiskarlag, a fishermen's interest organization that, among other things, has representatives at the Reguleringsrådet (Directorate of Fisheries, Norway). It is one of two large parallel interest organizations, which in fact are the result of a conflict between small- and large-scale fishing interests. Norges Fiskarlag was previously the organizing group for all fishermen, but was split into two groups – the original organization and Norges Kystfiskarlag – 'because so many of the small boats were unhappy with the policy of Norges Fiskarlag and felt that Norges Fiskarlag focused too much on the large boats … [they] set up a Kystfiskarlag of their own' (Directorate of Fisheries, Norway).[66] Norges Fiskarlag, which today represents large-scale interests, such as the sea fleet and shipping, is the politically stronger organization (Coast Fisher III, Norway):

> *Fiskarlaget [Norges Fiskarlag] has had a very, very strong influence on Norwegian fisheries policy for a very long time and now … the policies of the authorities [are] so very centred on having large-scale practices … larger and larger units because they say that this is what is most efficient. So now both the authorities' policy and Fiskarlaget's policy are united in this wish for larger units.* (Fishing Interest Organization, Norway; cf. Coast Fisher III, Norway)[67]

Some interviewees here noted that Norges Fiskarlag does not represent them, either democratically or with respect to their interests, and that the organization wields an influence that is detrimental to northern counties (Saami Parliament, Fishing, Norway; Fishing Interest Organization, Norway). Thus some of the interviewees noted that they did not see any reason to maintain membership of the organization:

'[The Fiskarlaget here in Tana] is dying' (Coast Fisher II, Norway);[68] 'They have never done anything for us fjord fishers' (Coast Fisher I, Norway).[69]

Another organization discussed by interviewees was the Norwegian Raw Fish Organization (Norges Råfisklag), which is a fishing sector interest organization that oversees that fish catches correspond with quotas and that sets the minimum fish price. It and other smaller-sales organizations were created by state legislation under the Raw Fish Act, which prescribes that all fish distribution should take place through sales organizations, because sales organizations can be required to organize fish reception points or fish transport to reception points even for remote areas (Directorate of Fisheries, Norway). The discussions about the Norwegian Raw Fish Organization centred on the stakeholders' perception that the organization constitutes a near monopoly on the buying and selling of fish, with a yearly turnover of some NOK 4 billion (Directorate of Fisheries, Norway). However, if fish reception were to take place to some degree directly between individual buyers and sellers through improved technology and better means for preserving fresh fish, 'that would have the result [that] the Norwegian Raw Fish Organization as a sales organization could not work as it has thus far … [where] the local fish buyer has had a monopoly … that is not the case anymore' (Saami Parliament, Fishing, Norway).[70] Given its position as a market actor, the Norwegian Raw Fish Organization was also seen as a potential topic of discussion if Norway were to become an EU member: 'We have had continuous adaptations to the EU through the EES [European Economic Space], so you would think that this would be something that would be taken up there' (Directorate of Fisheries, Norway).[71] However, the criticism of this organization was more limited than that of organizations and policies that were seen as influencing the present quota allocation.

Finally, interviewees noted that financial interests have a large impact on the fishing sector, as they are able to control which actors gain access to capital. One interviewee noted that since a 90-foot boat has loans for 'perhaps not a hundred million but several million NOK, NOK 40–50 million, it is a certainty that there are large bank interests in the whole thing' (Coast Fisher II, Norway):[72]

> *It is the banks that run things in reality, because the larger the loans you have … the banks signal to the Department that 'we must change this a bit so that they will survive in that group' … 90-foot [vessels] – that was the group that the banks supported. It is not only the banks but also the Fiskarlaget that have something to say about it.* (Coast Fisher II, Norway)[73]

The state funding body Innovasjon Norge was also criticized in this connection, because it was seen as imposing requirements as high as those imposed by the banks: 'Earlier they did not place market demands and the same demands on profitability, but on loans and so forth, and also on support, Innovasjon Norge places close to the same demands as the banks' (Finnmarks Fiskarlag).[74] On balance, interviewees directed their criticism at a variety of actors, the common objection being that these actors represent large-scale interests that appear to limit the allocation of quotas for fishing in Finnmark.

5.3.4. The impact of international norms and the international and EU levels

Generally, interviewees saw northern Norwegian fisheries as being regulated to a large extent by state decisions in committees with larger interest groups, and only somewhat influenced by local and regional interests. International norms regarding indigenous rights were one of the sources of authority that interviewees referred to in their aim to secure rights for Finnmark. ILO Convention No 169 on indigenous peoples' rights, ratified by Norway, was considered one particular means for influence:

> *Northern Norway – especially the fjords but partly coasts as well – these are Saami areas and that also means that the indigenous peoples' rights that are set out in ILO Convention No 169, for instance, [are valid]. Even though there has been a debate about whether the convention is valid for lake and sea areas, most legal experts are probably in agreement that it is.* (Fishing Interest Organization, Norway)[75]

Thus some interviewees stated that the inclusion of these rights in national legislation, which they felt had not yet taken place to the extent that they wished, would support the region vis-à-vis the state: 'One has [in Norway] ... supported what conventions there are to safeguard the interests of indigenous peoples and [other] minorities, but the formal legal and regulation framework has been entirely disregarded' (Fishing Interest Organization, Norway).[76] Yet interviewees did not discuss the ILO convention much more than this, possibly as it was perceived as having a limited effect on decision-makers in practice (Coast Fisher I, Norway). The practical implications of the convention with regard to both land use and fishing were, however, much more intensively discussed in the regional papers, in particular the local Saami newspaper *Ságat*. The thrust of the discussions was that the convention offers a possibility to defend local rights against the state.[77]

The interviewees also mentioned effects on local fishing from other international organizations, although they were less concerned with these than with the complexes of actors at the national level. The decisions on Norwegian state quotas are taken internationally by the International Council for Exploration of the Seas (ICES), based on research on sustainable catch levels and after hearing member countries. Many of the important quota allocation decisions for the northern Norwegian fishers are made in the Norwegian–Russian Fisheries Commission (Norske–Russiske Fiskerikommisjonen). These processes were briefly mentioned by actors. For example, one interviewee commented on the ICES as follows: 'It is possible that it is not entirely 100 per cent right but in any case the actors in it – the scientific basis –are naturally experts in marine research in each country' (Directorate of Fisheries, Norway).[78] With regard to decisions made in the Norwegian–Russian Fisheries Commission, one interviewee observed that he saw the same representation patterns there as in the distribution of national quotas: 'It is Norges Fiskarlag and the trawler industry which is at work there' (Saami Parliament, Fishing, Norway).[79]

Finally, one international impact mentioned was that of EU regulations on hygiene requirements for stockfish production and fish reception points and on EU customs barriers to Norway (which is an associated state but not an EU member). The comments thus pertained to regulations that directly affected interviewees (Saami Parliament, Fishing, Norway; Coast Fisher III, Norway): 'All of these fish farming units on the coast must invest enormous amounts in order to meet these EU demands. Otherwise they will not buy one single fish' (Coast Fisher I, Norway).[80] EU regulations were thus at times seen as positive, at times negative, depending on the costs they impose on local actors. With regard to customs barriers, interviewees noted that the EU customs barrier may contribute to the limited fish processing in Finnmark: 'Of course in connection with the EU the case of customs barriers on processed fish is … a factor that contributes to the fish being transported unprocessed out of the Finnmark region' (Fishing Interest Organization, Norway).[81] Customs barriers to the EU may be one reason why Japan and Russia rather than the EU are markets of particular interest and why the degree of processing in products sold to the EU – affecting the degree of possible processing in the region – is limited. One interviewee also noted that any possible entry of Norway into the EU would impact fisheries policy through, for instance, changing how national quotas are distributed (Coast Fisher III, Norway).

5.3.5. The governance network perceived by local small-scale fishing

On balance, the governance network where fisheries are concerned can be seen as a rather complex system. The major focus among interviewees was on the actors whom they associated with the national-level decision-making complexes. National decision-making was seen as comprising the state together with strong, large-scale fishing actors operating nationally, such as Reguleringsrådet and Norges Fiskarlag. These were the actors with whom the local small-scale fishing industry – supported by the municipality, the county and especially the Saami Parliament – saw themselves as being in conflict. In the eyes of the interviewees, there is thus a pronounced division of policy orientations between local and regional actors on the one hand, and larger-scale interests on the other. These divisions have separate support mechanisms and different policy aims. International norms and decision-making were discussed but emphasized less by interviewees than the national level, and were primarily mentioned for their value in exerting

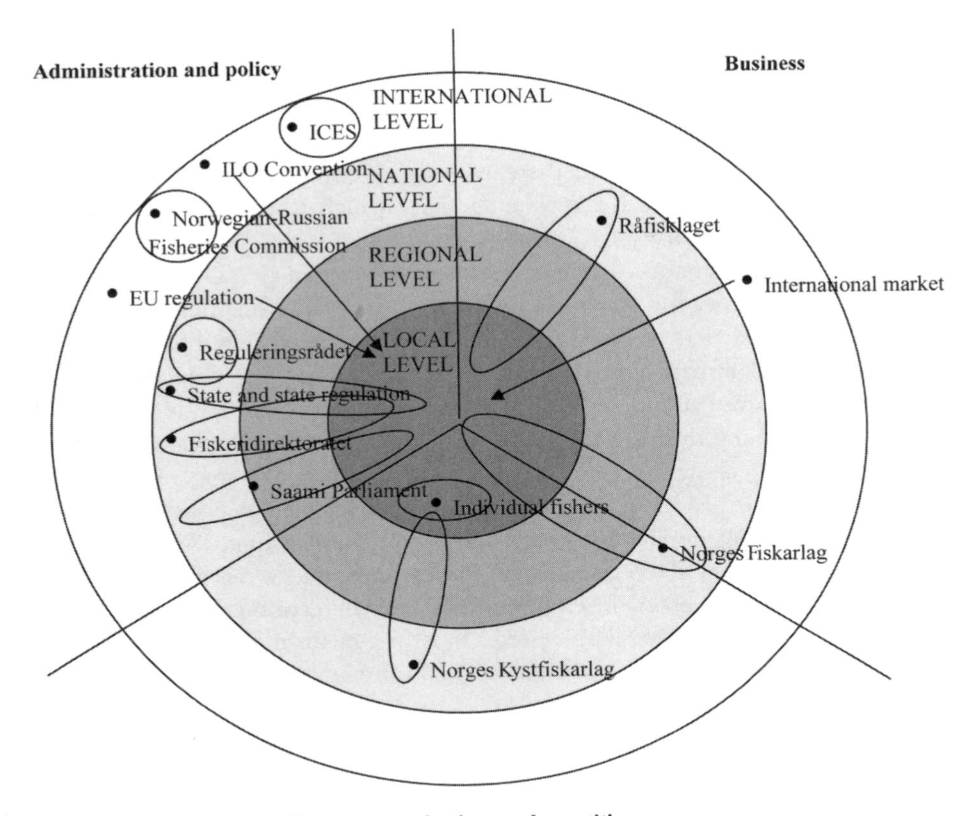

Figure 5.1 *Multi-level governance of fishing*

The diagram is divided into three sectors corresponding to the main groups of actors (administration and policy, business, and interest organizations and practitioners). Four levels of organization are distinguished: local, regional, national and international. The figure shows the principal organizations active in each sector, with dots indicating the level at which each nominally operates. Ellipses reflect the scope of an organization's influence as perceived by interviewees and arrows indicate large-scale influence via the market or regulation.

influence on the national situation (such as ILO Convention No 169). These networks and the actors within them are illustrated in Figure 5.1, which shows the multi-governance network in fisheries as seen by local and regional interviewees. The networks of local fishing are, as illustrated in the figure, relatively weak compared to those of larger-scale interests (an additional reason for this being the relatively sparse and aging population in the area and diminishing recruitment to the fishing sector).

The very clear focus on local conflict with the national-level decision-making was also emphasized in the interviewees' perceptions that the challenges to fishing are internal rather than external to the sector. One of the few outside conflicts with fisheries that was mentioned in general terms was that with the oil industry, which had recently expanded in the sea off northern Norway (Directorate of Fisheries, Norway; Finnmarks Fiskarlag). The only other interest group that was mentioned outside of the fisheries themselves were environmental organizations, and these mostly in general terms (Coast Fisher II, Norway). Competition for resources with external actors was not, as a rule, brought up.

There was also no mention in the interviews of conflicts at the local level, for example ethnic conflicts between Saami and non-Saami over resources. Local interviewees thus showed a relatively strong local and regional unity against large-scale interests in the protection of regional rights. This could be attributed to the policy adopted by local and indigenous organizations of not differentiating support for small-scale fishermen in the areas according to ethnicity, for example.

All in all, local small-scale fishing constitutes an actor that sees itself as relatively vulnerable within the Norwegian fisheries sector in comparison with larger interests, but has had some success in developing itself as a united regional interest in order to counteract this status. In the eyes of the interviewees, the role of the state was becoming more prominent, not diminishing over the period of time they described. Interviewees considered the distribution of quotas and quota policy as the major area for action, a perception emphasizing the nature of the governance network and the importance of regulation for Norwegian fisheries. As conflicts largely centred on the role of the state in allocating resources, the impact of international and EU-level regulation on fisheries were not emphasized,

Table 5.1 *Main perceived changes, adaptations and limits to adaptation in fishing*

Changes in the governance framework (economic and political changes)	Adaptations to perceived changes	Limitations on adaptive capacity
• Decreases in combination subsistence • Costs of fishing increased: more capital-intensive and technology-intensive • More large-scale: restructuring of fishing sector • Changes in processing techniques and products of processing • Institution of quota system • King crab invasion • Increase in competition • Increased diversification, e.g. into tourism • Limited recruitment and fewer active fishermen	• Political unity and policy development in region; suggestion of collective quota on regional basis, with an emphasis on indigenous and local rights • New types of processing (such as fresh fish) • King crab fishing • Fish farming • Shorter, more intensive fishing season	• Difficulty to obtain bank loans for small-scale fishing investment; difficult terms for loan repayment • Quota distribution system and policy towards small-boat decommissioning; effects from large-scale orientation that further limit infrastructure and tax revenues in the area • State's precarious position in relation to the Saami • Limited influence of small-scale fishing in lobbying and power structures

although potential means to influence this role, such as ILO Convention No 169, were mentioned.

5.4. Environmental and climate change in fishing

As seen above, interviewees focused on regulation (quotas) rather than biological production (amount of fish) as the factor directly determining catch levels. As a result, they placed comparatively little emphasis on environmental conditions and potential environmental change. Seasonal and other changes related to climate change were to a large extent perceived within the framework of a large natural variation in populations, which seem to be stronger in fishing than in reindeer herding and forestry, as fishing relies upon a freely migratory resource (Coast Fisher I, Norway; Coast Fisher III, Norway; Coast Fisher II, Norway; Salmon Fisher, Norway). Many of the interviewees focused their comments regarding the natural environment on what they saw as the most pressing matters of present environmental impacts – the king crab invasion in northern Norway and the impact of sea urchins and seals on fishing. They also noted that most adaptation to change would have to take place through individual economic adaptations within the boundaries set by the regulative system, such as investments in a quota or technology, which small-scale fishermen may sometimes find difficult to make.

5.4.1. Environmental influences on fish populations

Perhaps the most serious environmental problem perceived by fishermen was the king crab. Although it has a strong, positive economic effect for fishers who obtain a quota, the king crab also has a considerable impact on the environment: 'Researchers say that the crab does not have an effect on flora and fauna in the sea, but … what I and others have noted is what we call "the sea bed fauna" … has disappeared' (Coast Fisher III, Norway).[82] This is seen as a result of the crab's feeding patterns, which empty out areas to the point that the population disturbances through the food chain have now started to affect the crab itself. This prompted concern among interviewees (County Council, Norway) about the impacts for fishing and, above all, the decline in cod catches that the crab invasion is seen as causing: 'It [is] not easy to say whether there will be any fish here in ten years, because earlier there was very much of this small cod … but now you need to be very knowledgeable about the area to get small fish at all in the summer' (Coast Fisher III, Norway).[83] 'If [small cod] disappears then it will not take much longer until the [rest of the] fauna disappears and then other species will disappear as well' (Coast Fisher II, Norway; cf. Norwegian Raw Fish Organisation).[84]

An additional impact on sea floor and fish populations on the coast is the increase in sea urchins: 'It is not only the crab's fault that there is little fish, because we also have this sea urchin invasion that started perhaps 20 years ago; the sea urchins have grazed the sea floor bare of flora' (Coast Fisher III, Norway).[85] 'As a result, the small fish do not have many places to hide ... they are eaten by great cormorants and other birds and by seals and otters' (Coast Fisher I, Norway).[86]

As a result of the crab and sea urchin disturbance, interviewees worried that long-term changes may occur. Indeed some can already be seen as altering migration patterns: 'Some of the fish species seem to have changed their migration pattern because the larger saithe does not come into the fjords here like it did earlier and there is not any cod either in the summer' (Coast Fisher III, Norway).[87] Relatively complex interactions between different fish species, their prey and feeding grounds have been observed. For example, as a result of the changes:

> *The flounder has disappeared and the wolf-fish has disappeared without my having any good theory why they are gone. The seal cleaned the plate, but that was such a long time ago that the species should have re-established themselves, but it might have something to do with this crab, at least where the wolf-fish is concerned, because it spawns in a relatively limited area and in shallow water where the crabs eat the roe.* (Coast Fisher III, Norway)[88]

The large seal population is considered a disturbance, and one that has historically been limited through hunting. Seals are seen as consuming part of the potential catch: 'The local seals that we have, they require 5–6 kilos of fish per day to survive and if there are some thousand animals in the area that means a lot of fish' (Coast Fisher II, Norway).[89] Some interviewees criticized their not being able to limit seal populations through hunting: 'Sealskin was expensive, but now it is seen as illegitimate. The holy cow of the Department of Fisheries is the seal' (Coast Fisher I, Norway).[90] The criticism here was aimed at the regulatory and implementing bodies and at central assessments of seal levels: '[The Department of Fisheries] says that there are not as many as the fishermen say' (Coast Fisher I, Norway; cf. Local Government I, Norway).[91] Interviewees were also critical of the limited management of the environmental changes. Speaking about the degree of these disturbances, one interviewee even noted: 'It is an ecological disaster, but because we are situated here and not 20 kilometres outside Oslo, it is not a problem. Or if it is a problem it is not taken seriously' (Coast Fisher III, Norway).[92]

5.4.2. Climate change

In general, interviewees highlighted ongoing changes affecting the environment, noting that sea temperatures might rise and that this might have impacts on where fish populations occur and thus on their availability to small-boat fishers. Here, changes in individual species due to climate change could interact with fishing patterns to amplify variations. The impact of these changes on fishing would depend on whether economically important fish species were affected

(Saami Parliament, Fishing, Norway). Interviewees also noted that there already naturally exist 'large fluctuations in the populations of fish such as cod, saithe and haddock' (Saami Parliament, Fishing, Norway).[93] Today, 'through our regulation system, where we fish for particular year classes using particular techniques, [we] are increasing the fluctuations that occur in individual populations' (Saami Parliament, Fishing, Norway).[94] Climate change could further increase these variations in fish population, thereby further destabilizing those populations and the fisheries (Saami Parliament, Fishing, Norway). Adaptation to adjust to such variations and to decrease fluctuations could take place through, for instance, regulatory means for changing the timing of the quota in the year in order to target specific fish populations: 'you [could] reorganize things so you fish for the young fish populations that are from four to seven years old, which are so large here ... or [reorganize] the regulation year, which starts on 1 January, to start on 30 June' (Saami Parliament, Fishing, Norway).[95]

On the whole, impacts on fishing from climate change depend on the regulative framework, on existing fisheries equipment and resources deployed, and on the degree of change in – and interaction of this change with – natural and regulation-induced variations, as described below.

5.4.2.1. Autumn and winter

Many interviewees were concerned about the changes they had noticed in autumn and winter. Interviewees noted that in the last few years the weather had fluctuated beyond what they considered the normal range: 'Suddenly it starts to rain in the middle of the winter and is mild for 14 days. That is not normal' (Coast Fisher I, Norway).[96] 'The winters up here in Finnmark ... start to remind me more and more of the Lofoten winters' (Norges Kystfiskarlag).[97] Thus, there is 'mild weather in the middle of ... the winter, and long periods [of it]' (Norges Kystfiskarlag).[98] Interviewees also noted that winters had less snow, and that autumns had become longer (Salmon Fisher, Norway; Coast Fisher I, Norway; Coast Fisher III, Norway).

Some changes in autumn and winter could especially impact river fishing. Changed precipitation and temperature patterns, which could include rain in winter, were seen as impacting the location of fish populations and the relative length of the fishing seasons in rivers: 'Two years ago it rained and rained, and if it had continued to rain the ice on the River Tana would have gone' (Coast Fisher I, Norway).[99] Certain changes in the location and size of the salmon populations in the river, possibly related to changing precipitation and flood patterns, have already been noticed: 'This entire year has been extremely bad around all of Finnmark. There is no salmon anywhere' (Coast Fisher I, Norway).[100] One reason for this may have been the elevated water level in the river in winter: 'The spawning must succeed, and it may be that the flow in the river was so strong when the fish released its roe that it froze onto the ice and was destroyed' (Coast

Fisher I, Norway).[101] 'It is like the old men up along the Tana say – that it is better if there is little water in the river when the salmon spawn. If the river is high, the fish spawn on the gravel banks, which dry up during winter' (Coast Fisher I, Norway).[102] Water levels are thus emphasized in the case of salmon fishing (Coast Fisher I, Norway). Here too, however, natural variation makes it difficult to determine whether this is a change related to climate change: 'It may be that there are natural fluctuations and it has happened before that there have been bad salmon years' (Coast Fisher I, Norway).[103]

With regard to coastal fishing, interviewees noted that 'there has not been ice on the fjords anywhere, unlike 20–30 years ago when there was ice far out in the fjord here ... each winter' (Coast Fisher I, Norway).[104] According to the interviewee, fjords have been ice-free for some 15–20 years already, or even longer (Coast Fisher I, Norway; cf. Coast Fisher III, Norway). Direct impacts on fishing, however, were less frequently discussed, and interviewees often referred to scientific findings when commenting on trends: 'The temperature in the sea has risen by perhaps one or two degrees, that is what the researchers say ... the temperature is rather high compared to what is normal' (Coast Fisher I, Norway).[105] Interviewees also mentioned that changing winter conditions might make boats ice over and that unexpected storms could cause problems for fishermen already out on the sea. However, no increase in storms has been noted to date. On the contrary, 'in recent winters ... there has not been a single day with strong wind. We have escaped the worst winter storms now for several years in a row' (Coast Fisher I, Norway).[106]

The actual impact of weather during particular times may also be limited: 'Many fishermen, they do not get stressed out like that in January and February, the darkest and hardest time of the year, because they know that they will catch their quota anyway' (The Norwegian Raw Fish Organization).[107] The regulatory framework with its small quotas may thus in the long run make questions of access due to weather conditions less important for fishermen: 'What people talk about is incurring the least possible expense from catching the fish ... You look at oil consumption, what you use in oil to get those ... tons [of fish brought in]' (The Norwegian Raw Fish Organization).[108]

5.4.2.2. Spring and summer

In general, interviewees noted that spring, too, started early, with 'early warmth in the spring, such as here on 1 May or at least 15 May, when it was up to 20 degrees. That is entirely ... abnormal and last year there was something similar' (Coast Fisher III, Norway).[109] The warmer temperatures are seen as changing fish migration patterns: 'There are warmer temperatures ... The last few years [there] have been very large fish here, fish that have normally gone on to the Lofotens to spawn stop here off of Finnmark. We notice that there is very much roe in the fish ... delivered in Finnmark by the fishermen' (The Norwegian Raw Fish Organization).[110]

Other interviewees noted, similarly, that 'the cod have begun spawning off Troms' (Norges Kystfiskarlag).[111] Such a change towards warmer temperatures could be positive: 'Warmer water and such will give you quicker production. That would be directly positive for many things. Cod would grow better and spawn. The plankton would grow and you would get a good nutritional [supplement]' (Finnmarks Fiskarlag).[112] However, the same interviewee noted, 'This is not the way it should be; it is probably not positive in the long term. There will probably be a backlash in one or the other way' (Finnmarks Fiskarlag).[113]

However, even if rising temperatures did result in there being more fish, this would not necessarily have a direct positive effect on fishing quotas and the economic situation of fishermen and their livelihood: '[That there will be more fish] might well happen, but if … the quota increases with it we do not know … if it will reach the fisherman in the boat' (The Norwegian Raw Fish Organization).[114]

It was also noted that increasing temperatures and precipitation may increase algae and pests. One actor noted that, 'We have actually had algae for a couple, three summers now. The sea becomes green, but there has not been any pronounced problem' (Norges Kystfiskarlag).[115] On the other hand, some other potential problems were observed. If the prevalence of fish parasites, even those that do not reduce quality, increases, this may affect the market: 'It can cause a great deal of damage in the market and in getting people to eat fish and such … one [time] in the 1990s … the saithe had … *kveiste* [parasites] and in just one day the entire German market dried up' (The Norwegian Raw Fish Organization).[116]

A more severe impact, however, would be a change in the geographical distribution of fish species. This prospect in particular illustrates the regulative character of environmental systems in fishing. If, for instance, the more southern species of mackerel were to become more common off the northern Norwegian coast, it would not necessarily be something that northern Norwegian fishermen could benefit from: 'We have not needed a fleet for fishing for mackerel and herring … That requires capital and all of that, also permits … you need to have a mackerel quota. You need to have a thick wallet if you are going into herring and mackerel fishing' (The Norwegian Raw Fish Organization).[117] To some extent, indications of change – such as mackerel migration north – can already be seen: 'The mackerel is coming further and further north. We already have it up in Vestfjorden. They never talked about mackerel when I lived there in the 1970s, not that I know of in any case' (Norges Kystfiskarlag).[118]

An additional problem would be if the geographic distribution of fish species overall were to change, with fishing becoming less accessible to coast-bound fishermen. One actor summarized this by saying, 'If the fish do not come to the coast because of the temperature, the small-scale fishing fleet will not have a chance' (Finnmarks Fiskarlag).[119] This is partly a result of the risks for smaller boats posed by storms at sea and partly of increased costs of fuel and other operations. While fishing today takes place throughout the year, peak times for fishing are in spring (quotas are also set at the end of the year, before the new fishing season, meaning that fishers can start fishing with a new quota in spring). At that time, northern winds and high winds may prevent fishing (County Council, Norway): 'The higher the temperature ... the more unsettled the weather' (The Norwegian Raw Fish Organization).[120] However, as noted above, bad fishing weather during the season may have a relatively limited impact on fishing levels: 'Now the quotas are so small, 99 per cent of the fishermen will catch their quota anyway' (The Norwegian Raw Fish Organization).[121] Consequently:

> *If one speaks about how fishermen adapt ... there are basically two strategies. One is that most of the younger fishermen will remain in the business. They will buy whatever it takes, that is they invest in quotas to get a better starting position. Then there are those who are ... I do not know if it is a strategy or a lack of one: they do nothing, they count on making a living during the time they have remaining [till retirement].* (Finnmarks Fiskarlag)[122]

There may be effects on different fishing vessel sizes depending on how well they deal with strong wind. The small boats that are typical of the Finnmark area are especially restricted by unfavourable weather conditions: 'The Finnmark fleet is not built to go out to sea. It is a coastal fleet, adapted for the coast' (The Norwegian Raw Fish Organization; cf. Local Government I, Norway).[123] On the whole, however, technological changes have meant that fishing boats can cope better with bad weather than before (County Council, Norway). Here there are differences between fjord and sea conditions, with bad weather being far more serious on the open sea: 'You get to the point where [going to sea] is a physical impossibility' (County Council, Norway).[124]

Should there be a shift in the geographic distribution of fish, a small-boat fisherman might lose out by not having the equipment for fishing out at sea: 'You do not have ... the right-sized equipment for it, the depth and current are larger. You ... do not have ... this tying equipment for nets or nets that are the right size for it. ... You cannot change over so easily; it is either one or the other' (Finnmarks Fiskarlag).[125] This situation, however, is partly a result of regulatory decisions. Fishing farther out from the coast than some ten nautical miles requires a certificate (The Norwegian Raw Fish Organization; Norges Kystfiskarlag). Moreover, the fishing industry in Finnmark may not have the capacity to deal with new species. For instance: 'You have a fillet industry that is built on cod, with machines built for cod. They cannot cut mackerel, or other fish species' (Finnmarks Fiskarlag).[126] What is more, the area does not have a tradition for, for instance, smaller-scale mackerel processing: 'We do not have the same traditions ... those who have hung and dried fish and transported it in the Lofotens for generations ... they have technologies for that process ... and have contacts with the importers in Italy' (Finnmarks Fiskarlag).[127] A change in fish species would thus require not only different equipment but different sales networks as well. Warmer weather could also threaten existing processing methods. Interviewees noted that there had been some impacts on the quality of dried fish: 'It is probably a result of temperature. To get perfect quality you are very dependent on temperature ... it can go sour quickly if it becomes mild and such' (The Norwegian Raw Fish Organization; cf. Finnmarks Fiskarlag).[128]

Impacts on river fishing can also be identified. A good salmon year would have little wind and relatively cold, not too high water (Salmon Fisher, Norway): 'In spring, when we have much precipitation and the rivers are high, the fishing is very bad. There is a period during the spring when there is much fresh water in the sea and a very large production of plankton. It is a very bad fishing period' (The Norwegian Raw Fish Organization).[129] Warm water temperatures were seen as resulting in low salmon catches: 'It is the worst [year I have had]. I have never caught so few salmon' (Salmon Fisher, Norway).[130] This was seen as a result of 'the water being up to 20 degrees in the river ... If the water is continuously becoming warmer the salmon will seek out the deep areas' (Salmon Fisher, Norway).[131] Salmon catches have decreased during the recent warm years: 'One can look some three years back when the catch started to decline, but it probably has something to do with the water having become so warm' (Salmon Fisher, Norway).[132] Additionally, river fishers noted the impacts of limited equipment, which prevent them from following fish resources. For instance, one interviewee gave the following reason why people fish in the river in Tana, although the sea is nearby: 'It is without doubt [because] the fish go deeper [in the sea or fjord] in the

Table 5.2 *Main emphasized climate change sensitivities and adaptations*

Projected climatic change	Local sensitivities and adaptations
Warmer water temperatures, causing changes and geographical shifts – mainly northwards – in species, with possible increases in fish biomass and growth	Changes in individual species such as mackerel; new species require other quota and equipment (however, local industry is adapted to certain species and does not have traditions for other fish production)
	Changed migration patterns and geographic distribution result in decrease in availability to small-boat fishermen, as they cannot go to sea (due to equipment and regulatory restrictions)
	Quicker production may cause algae and pests to increase, which may impact markets; quota limits mean that fishermen may not necessarily fish more even if there are more fish
Extensive variations in weather	Impact of weather events such as storms are limited, as quota levels can be caught during a relatively short period which can be adjusted
Warmer summers	Increased warmth may impact river fishing and spawning negatively
Changes in precipitation	Increased rain may negatively impact river fishing and spawning

winter in January and especially in February and that many do not have boats to go further out into the Tana fjord' (Coast Fisher II, Norway).[133]

In sum, interviewees emphasized throughout the large variability in fishing from year to year and the difficulty of assessing whether the changes they observed are part of longer-term trends or not. The changes that were uppermost in the fishermen's minds were the distinct environmental changes (in addition to specified climate change impacts) that could be seen in the fjords and could in part be attributed to the long-term king crab invasion. Interviewees were thus concerned with impacts on sea floor flora and fauna that may limit future fish populations. Changes in geographical fish distribution – especially given the limited adaptability to changes in fish distribution and species – were otherwise seen as having the largest potential impact on fisheries.

More general changes across the freeze–thaw line (onset of spring varying or spring coming earlier, milder winters with possible thaws, and later autumns) have been observed but are not seen as having a major impact on fishing.

5.5. Conclusion: Vulnerability and adaptive capacity

The vulnerabilities for fishing in the area can largely be associated with the present situation and governance framework, in which fishermen feel that their employment and living situation are threatened. The fisheries sector is based on a finite resource and its maintenance, as well as on the distribution of the resource in the form of quotas to different fishers. The adaptive capacity of the sector is constrained by this situation, and is a sum of adaptations through time as well as of the resources and governance framework that presently constrain what adaptations can take place. For local fishing, agency is to a large degree determined by regulations at higher decision-making levels, especially the state and associated decision-making or other influential bodies, which limit the adaptations that can take place without coordination across multiple levels to support local adaptation.

5.5.1. Individual adaptations

Fishermen noted that small-scale fishing has historically adapted by seeking out accessible areas and adopting viable practices. These adaptations are regularly small-scale and fall within a framework determined by economic resources and regulations. Given that the quota system precludes increasing the catch, economic adaptations for individual fishermen are principally associated with the market, one example being fresh fish sales to raise income per catch unit. Other small-scale adaptations mentioned by interviewees include combination subsistence and diversification into tourism or fish farming. These forms of adaptation in fisheries, which to a large extent rely on the individual fisherman producing fresh fish or diversifying into additional occupations, change the focus of adaptation in that the individual must rely on himself or herself rather than the societal network. If dissatisfied with the distribution within the quota system, actors in the sector may increasingly rely upon individual adaptations, perhaps complemented by political protest. Adaptation is also limited by a lack of resources for funding reorientation and adaptation, for instance in the form of fish farming, loans for boats or tourism-related activities.

The economic adaptations that each fisherman can undertake are also limited in that the fisheries operate in what is inherently a very changeable market. Price levels play a large part in determining whether fishing is profitable in the short term. Today, for instance, when the price and demand for wild salmon is lower, such fishing is less economically attractive to fishermen. However, fishermen noted that the market and getting fish sold do not entail difficulties. There may also be new markets developing, such as Russia, for which Norway would have a competitive advantage in transportation times compared with many other countries; competitive advantage can be anticipated in both fresh fish production and in geographical proximity. A possible restriction on this market development lies in the relatively high Norwegian fish prices.

The main focus regarding individual adaptation was diversification and adaptation to the market. Vulnerability to change and within the current situation was largely seen as varying depending on the fisherman's access to quotas and especially crab quotas. Many of the economic adaptations are thus interrelated with political as well as resource factors, although, due to the quota mechanism, increases in the resource base were not considered directly translatable into increased resources for the fisherman. As so many of the available economic adaptations are determined through regulatory limits and the political system, it is difficult to distinguish clearly between political and economic adaptations. The fact that individual economic actions are constrained by the regulatory framework is particularly apparent in fishing, given the clear impacts of the quota system.

5.5.2. Larger-scale adaptation within the governance network

Political-level adaptation and regulatory change were the areas that were by far most frequently criticized in the interviews.[134] Current situations may thus be seen as a result of situations and adaptations over time within a complex web of governance that includes government, administration and interest organizations (as well as, to some extent, the scientific basis of resource decisions).

According to interviewees, the main political or regulative adaptations should take place as part of national-level changes in regulations affording small-scale fishers in northern Norway larger economic leeway. This was discussed in conjunction with demands for changes in the *Kondemneringsordningen* and quota distribution, which could be effected by changing decision-making structures or the *Finnmarksloven*. Interviewees were in general displeased with the national regulations and would prefer a system that made it easier for small-scale fishers to obtain group 1 and crab quotas, that kept quotas in the region, and supported the landing of fish in local and regional ports. For instance, interviewees suggested that the distribution of quotas to fishermen should to some extent take place regionally, or at least with the region playing a larger role in quota management.

Regarding larger-scale political adaptations, interviewees also suggested that regulations could be changed to provide additional structural support for individual economic adaptations and for recruitment in order to maintain a viable small-scale fishing population. However, the stakeholders interviewed in the fishing sector noted that they felt they had comparatively little influence on this decision-making system. Much of the decision-making takes place in national-level bodies, which interviewees largely saw as driven by large-scale fishing interests, such as industry and owners of large fishing vessels, as well as by financing interests. Interviewees generally felt that they have few possibilities for providing feedback to this system and would prefer more regional decision-making rights. They emphasized that there is overcapacity in fishing today and that many of the large fishing vessel classes, such as trawlers, are so efficient in harvesting fish that the entire Norwegian quota could be caught by some 20 trawlers working the year round, which would have tremendous adverse impacts on coastal population in term of employment. Interviewees thus posed the question of whether the present management is sustainable and considered local small-scale fishing to be preferable politically.

5.5.3. Adaptations to climate change

Regarding adaptations to climate or environmental changes at large, interviewees noted that fisheries as such have a large natural variation to which adjustments are made within the political and economic framework. No particular adjustments to climatic changes were being considered by interviewees at the time the study was undertaken – among other things because the regulatory system is seen as having a larger direct influence on fishing rights than the environment does. Any change or adaptation in fishing would need to take place through the quota system and other regulations, such as by granting quotas to catch other species or certificates to go further out to sea. Technology also limits adaptive capacity, as adaptations such as the above would require large and costly reinvestment for both individual fishermen and the industry. Where climatic conditions such as storms are concerned, the vulnerability of fishing is limited or obscured by its thorough integration and practice within a regulative system, as the limited quotas can be caught in relatively short times when the expenses are lowest, in other words when there is less chance of storms, high winds, precipitation and cold weather. Many potential adaptations to climate change would thus require raising actual adaptive capacity for the local level through action within the political sphere, aimed for example at quota distribution. This illustrates the close interlinkage between economic and political systems in fishing, where the regulative framework very clearly sets the boundaries on available economic adaptations. For instance, one adaptation to cope with large changes in fish stocks and their distribution as a result of climate change could be regulation to start the quota year in June rather than in January. The timing of fishing over the year could be seen as a partial adaptation that might be less difficult to effect than some of the other suggested modifications.

5.6. The vulnerability of local fishing

On balance, local coast fishing, as well as river fishing, may be considered relatively vulnerable, particularly in comparison with other types of fishing. Local vulnerability is differentiated at the individual level depending on a fisherman's access to the basic resource through quotas; this access was seen by interviewees as generally delimited through a policy framework influenced by a complex of large-scale financial and economic interests, especially at the national level. Environmental feedback, economic feedback and the increasing competitive demands of the world fish market are filtered through this regulative system.

Although the issue has not been investigated in this work, at least one interviewee suggested that these problems are common to small-scale fishing, where they can be attributed to increasing rationalization and efforts to make fishing a large-scale industry:

> *The coast fishermen have the same problems everywhere, that is, all around the Atlantic Ocean. In Portugal, Spain, the Faeroes or Iceland, they have exactly the same problems, and are in decline because of large-scale capital interests and big business.* (Norges Kystfiskarlag)[135]

6
Conclusion

6.1. Economic, political and environmental changes

The foregoing chapters reveal a broad spectrum of changes and adaptations in the sectors studied. Although it has examined three different sectors, the research has revealed some strong similarities, including trends that span different countries and illustrate structural change in which globalization could be seen as a common impact. This chapter will summarize the more pronounced factors of change that have emerged in the study, in other words those spanning sectors and areas. In these instances, as interviewees in fishing suggested, trends may be relevant beyond a particular region or sector and reflect broader globalizing tendencies that can be expected to impact vulnerability in similar ways elsewhere (depending on the local structure, including, for instance, the degree of marginalization in the area). Examples of trends that could span larger areas include stakeholders' perception of a growing emphasis on the market and economic conditions; an increased delocalization, or uncoupling of production decisions from the local level; adaptations that limit vulnerability by removing sectoral practices further from the natural environment and natural variability; and the enhanced capacities of systems to differentiate vulnerability among different groups.

The study also illustrates the utility of viewing adaptations and adaptive capacity as determined within frameworks of governance. Economic adaptations within a market system are most often those that actors in the particular economic sector or livelihood are used to adopting individually, such as developing market niches and market networks, and changing the quality or type of product offered. The framework within which these adaptations are undertaken is set by an international market with globalizing tendencies and a political regulatory framework; together these comprise a governance structure that limits resource use and distributes privileges and support to different actors (for example by opening or closing certain avenues for adaptation, which sometimes makes short-term adjustments – or coping – viable over the longer term).

Localities may respond to such developments, which occur at largely national and international levels, in the form of an increasing local unity within the relevant sectors or by increasing local resource conflicts as a result of greater external demands, in other words by maximizing adaptive capacity within a multi-level framework in different ways. Accordingly, the chapter will also discuss whether local fragmentation is taking place as a result of resource conflicts, exacerbated by globalization and climate change tendencies, and possible differentiation among local actors into winners and losers in the processes of change. Here, the research reveals the different ways in which identity or ethnicity is interpreted in relation to resource rights, and how ethnicity can work to unite or divide people depending on its interpretation in the different case study areas. The sections below compare the changes in the governance framework described in previous chapters with the changes expected as a result of economic and political globalization; the conclusion drawn is that economic globalization in particular figures prominently in the areas, whereas in the case of political globalization the picture is more fragmented, exhibiting tendencies towards both globalization and the continued dominance of the state. Shifts in the pattern of decision-making can, however, be seen at different levels. Climate change, which may have a large impact on the areas, has the most direct effects on stakeholders in reindeer herding; stakeholders in forestry and fishing have perceived and anticipate more limited impacts. In some cases, this may be due to limited awareness, but in others it is due to the major role of regulatory systems in mediating and determining vulnerability to climate change.

6.1.1. Economic globalization?

The impact of factors commonly described as 'economic globalization' have been notable in the areas studied, including the internationalization of production and trade networks or commodity chains and a decoupling of these from the local level. This is reflected in the fact that changes in the viability of livelihoods was the stakeholders' main concern in all sectors. The local level may also be affected by state measures for dealing with globalization, such as the restructuring of government agencies, decreasing employment in state forestry or the supporting administration, and reductions in state spending overall, including fewer subsidies and the decentralization of government services. International economic competition has had a considerable impact in all the cases, as evidenced by an increased adaptation to the international market and an expansion of export operations over time. Additionally, the number of employees in the sectors studied has been reduced through rationalization, and company employment has been replaced to some extent by mobile labour, one example being the new class of contractors that has emerged in the forestry sector.

In forestry, one can see a relatively high number of economic restructuring factors, especially in relation to international production networks. Interviewees also generally described a marked increase in internationalization during their time of employment. Even small-scale enterprises considered participation in international networks and trade to be a survival factor. In reindeer herding, interviewees observed that the demands made on reindeer herding by the market economy have increased during the time that the herders have been active in the business. Herders have been forced to make production more efficient and to use supplementary feeding – not only to ensure themselves an income from herding but also to compensate for disturbances from other sectors that have reduced the amount of pasture land available. The meat market and price levels were the topics which reindeer herders brought up most frequently in the interviews. The most salient aspects of the issues that were discussed included the meat market surplus, the limited number of buyers and the need for more consumer-oriented measures, such as marketing reindeer meat in order to increase demand. Interviewees noted that the meat market is influenced by, among other things, competition from deer meat produced in New Zealand and EU taxation borders and regulations. These developments reveal a process of internationalization at work in reindeer herding that is different from what one might have expected in a relatively 'traditional' livelihood.

Similarly, in the case of fishing, interviewees described how the occupation has changed over time to become larger in scale and heavily regulated. The market and market mechanisms are largely taken for granted. For example, adaptation to changing market requirements, such as those leading to the relocation of fish production units to areas with lower costs, were mainly discussed in terms of the need to adapt rather than as causes for political action (in contrast to quota setting). Among the ongoing adaptations discussed were different types of fish processing in the form of fresh fish production. This situation indicates the relatively extensive integration of fisheries in the global market. Examples are the export of locally caught resources to China and Russia and the competition with these countries, among others, in the area of fish processing. The end result here may be that large-scale actors are able to reap the benefits of globalization and internationalization while strictly local actors suffer from reduced state support.

All in all, elements of economic globalization are readily discernible in the sectors. Taken together, these changes have increased vulnerabilities. One example of this trend is that the most important markets for local products are now international. This correlates with the high degree of internationalization of production and marketing discussed in the globalization literature. This internationalization holds true for wood products and fish in particular, but applies to reindeer herding as well, as evidenced by the large drop in Finnish prices when Norwegian buyers stopped importing Finnish reindeer meat.

Many of these more general changes have been described elsewhere, especially for forestry (Layton and Pashkevitch, 2000) and fishing (Nuttall, 2005), but have seldom been placed in the framework of adaptation. In instances such as these, changes in the market can enhance or reduce the value of adaptations; for example, in reindeer herding a lower market value for increased meat production will decrease the value of supplementary feeding. Although globalization may enhance adaptive capacity to some extent by providing access to larger markets and outlets for products among those best able to utilize the system, it disadvantages actors who cannot access or compete on those markets and renders communities in which many such actors live and work vulnerable. In other words, the vulnerability of local areas to a large extent lies in their dependence on and the uncertainty of the international (and domestic) markets. However, the factors that determine who can operate effectively within these systems are not purely structural, although structure – the size of a business or economies of scale – is crucial. For instance, actors who themselves develop international networks of buyers (such as one Swedish sawmill owner) are to some extent able to avoid problems of limited access to markets, buyers and internationalization and thus to improve their adaptive capacity.

This observation underscores the importance of examining local vulnerability within a multi-level and multi-impact context and of viewing not only general or structural factors (larger actors with scale advantages, embedded in buyer networks in a functioning market) but also more individual factors (such as entrepreneurship) that determine actors' vulnerability. However, given the focus on stakeholder knowledge and perceptions in the study, changes on a larger timescale than about a generation have not been considered. Such a perspective might have revealed extensive structural changes stemming from different causes that would have contextualized and provided a comparative perspective on both the magnitude of perceived change and the underlying factors and causes of change beyond those cited by stakeholders.

6.1.2. Political globalization? The pattern of governance and the role of traditional policy actors

Where political globalization is concerned, the pattern of change in the areas is not as clear as in the case of economic globalization. In the literature, political globalization is regularly seen as including a destabilization of the state in its control over economic means and devolution of decision-making powers to private actors, among others; the profound impact of changes in the international market discussed above is a salient example. However, one principal focus among the interviewees when discussing decision-making was regulation by the state. This could be seen especially in their comments on the impact of state ownership on forestry, but also in discussions of national fishery quotas.

Overall, the impacts of international political actors on the local and regional levels in these regions can often be considered limited in comparison to those of national governance networks. This suggests that the state remains a crucial actor at the local level and indeed that a large part of adaptive capacity is situated and determined at state level rather than at others. However, political globalization in the cases studied appeared as an increase in the importance of the international level, especially international norms, and in the indirect impacts of these norms transmitted through the state level. Political globalization can also be seen directly in the increasing impact of international norms at the local level. Norms such those found in ILO Convention No 169 were discussed as one means to influence the state where fisheries and reindeer herding were concerned, while for forestry the market-driven, non-governmental forest certification scheme addressing environmental and indigenous concerns (and developing regulative networks that sometimes imposed more stringent requirements than state legislation) was emphasized. That these instruments figured prominently in the minds of the interviewees reflects the international impacts on local resource debates that reconsider the relationship between the state and the people at large, between local and indigenous populations, and between government and the market as steering mechanisms. Finally, interviewees, especially in forestry, often regarded environmental protection processes, in particular Natura 2000, as having negative impacts on their access to primary resources. On the whole, the study illustrates that environmental and indigenous norms on the international level are concerns among a number of local actors.

The governance networks that interviewees described for the respective sectors illustrate multiple levels of control at multiple sites that reflect possible indirect impacts of political globalization. Most of the bodies seen as affecting local agency are governmental and administrative, most operate at the national level, and yet there is a substantial impact on local actors from the international level. Local areas were seen by interviewees as losing more and more of their decision-making rights to large companies – increasingly ones located outside the area – or to the national level, as in the case of fishing quotas. The number of regulative units (state administration/governance on different levels) has decreased, with fewer employees and fewer branch offices, and the integration of several municipalities into larger ones with a joint administration was discussed in many areas. These developments may remove the increasingly regulated administration even further from the local area, decreasing possibilities for direct democratic participation. It may be such a thinning of the state presence that has actors leaning more towards institutional norms beyond the state, such as forest certification or ILO Convention No 169, which accord them a certain measure of influence.

Thus, while the governance network that makes decisions affecting the sectors comprises numerous actors, many of the decisions are taken within economic and governmental regulative bodies which the local level can only influence to a limited extent and which it has little opportunity to give feedback to (a pattern suggested by Naess et al, 2005). This situation has considerable implications for local adaptive capacity, where potential and actual adaptations are constrained by higher levels and there are few possibilities to communicate adaptation needs to decision-makers. The discussions of international norms in the material can be regarded as sometimes uniting, sometimes dividing local groups depending on the impact of those norms on the stakeholders' livelihoods. The norms may affect resource access either directly through a convention or agreement or indirectly though their potential implications for the state regulation that governs such access.

Most of the interviewees indicated that there were limited possibilities for reducing vulnerabilities or increasing adaptive capacity through governance networks in the sectors. While local sectors may be part of major chains of interest, they do not have much influence on them. Local forestry, even if part of a relatively well-institutionalized and major sector with developed buyer and market networks, constitutes a rather dispersed and limited actor – or, rather, group of actors – greatly affected by wide-ranging national and international changes. The limited possibilities for diversification and the sector's heavy rationalization have increased vulnerability locally. Reindeer herding represents a rather thin network of interests, with few strong buyers and a limited demand for reindeer meat (except perhaps in the case of northern Norway, where the sector is of relatively large economic importance and the especially strong Saami element may serve to protect resource rights). As a result, reindeer herding has difficulty asserting itself against competing land uses (in contrast to forestry, for instance). It is here especially that international means to exert pressure on the state – the main regulatory body – have been emphasized, in particular ILO Convention No 169.

In the case of fishing, the implementation of the quota system was cited by interviewees as the greatest problem for small-scale fishing. They did not indicate any objection to the system itself but opposed the national distribution of quotas between large- and small-scale fishing in what is largely an intra-sectoral conflict. In the northern case study area, however, small-scale fishing actors utilize ethnic markers to a large extent in order to promote rights for the region as a whole; they form a relatively united front in this effort and increase their leverage as a political interest group despite their relatively limited scale and role in comparison with larger groups. The role of the state system is pronounced here: as in forestry, the political and distributive role of the market in determining production variables is more readily accepted as a requirement for remaining competitive, while the political role of the state and state limitation and distribution of resources is contested.

6.1.3. Interaction between economic and political globalization: Governance by the state and the market

On the whole, this study illustrates that general adaptive capacity should be analysed in terms of the principal impacts of ongoing and anticipated economic, political and social changes on stakeholders' livelihoods. There are also clear interlinkages between economic and political systems, particularly regarding employment, within broad, multi-actor and multi-level systems of governance for the sectors and areas. Interviewees within the sectors and areas display a very similar focus on the need for maintaining employment levels in the region, an aspiration that is expressed both politically and economically. That both of these channels are used illustrates that regulation of activities takes place not only at the domestic level but also in large, market-driven systems where responsibility is difficult to assign and 'complaint possibilities' are few. The sectors also have some similar problems, among them financing and sector-internal conflicts between large- and small-scale actors. Interviewees in the sectors revealed a preference for individual economic adaptations, as these are the ones over which they have the most control. A problem common to these adaptations is that the actors cannot control the most important factor, the market system (for example reindeer meat prices). Here, technological adaptations, which constitute a major trend over time, may support individual actors and the sectors at large economically by increasing efficiency and enabling rationalization; then again, the adaptations, in making resource outtake more effective, may also add to the existing range of resource problems and over time even exacerbate economic problems in the areas, given that additional rationalization may further limit employment. The economic support provided to the areas studied, for example through the state and through a domestic and regional focus on financing by lending institutions, seems to have decreased, possibly due to the sectors' more limited role as employers in the region, possibly due to a shift in focus by the state regarding regional support. Forestry and fishing are sectors where access to capital or financing often determines whether an entrepreneur can continue operations and a lack of capital will restrict local entrepreneurship. Banks play a key role here, as they are the largest potential group of lenders; their policies illustrate the interlinkage of political and economic factors, with industries judged as less economically viable over time losing political support within a market system. What one can see here – in the case of both forestry and fishing – is the importance for adaptive capacity of state and private loan institutions and their willingness to support a particular locality and relevant sectors.

The interlinkage between economic and political globalization also appears in how the bottlenecks for each sector are located in different parts of the sectoral systems. For forestry, one important bottleneck is the supply of wood, especially local wood, which many smaller local companies consider crucial. For reindeer herding, a crucial bottleneck is the market's ability to absorb the reindeer meat that herders have been able to supply through individual adaptations. In the case of fishing, interviewees found a major stumbling block to be the distribution of quotas, which limit catches (in other words supply) and therefore potential income. All of these factors limit a company's potential economic performance through either market deficiencies or regulative, political measures. This trend parallels the development over time from relatively freely used resources to increased regulation, where the resources are no longer open-access commons. As already mentioned, regulation takes place through not only the governmental system but also market-directed frameworks such as certification. Restrictions are enforced, for example, through state systems such as fish quotas, limits on reindeer numbers (which in some instances approach quotas), forestry resource ownership, coordination with reindeer herders and certification. New market regulations also exist that further limit resource access; in principle, observance of these regulations is voluntary, but in practice – at least in some areas – it is a requirement for access to markets. One example of this is certification in forestry, which can be regarded as indicative of larger global trends (for instance, several eco-labelling systems for fishing are today in use).

Resources are thus increasingly regulated, through both formal regulation and market-based means, evidencing not only the importance of new governance measures but also the continued importance of the government (Hooghe and Marks, 2003). Interestingly, of these two forms, interviewees generally perceived market demands as a given, examples being forest certification or market changes resulting in the relocation of fish fillet production. Dissatisfaction was instead mainly targeted at governmental regulation, such as the state-managed forestry in Sweden and Finland or fishing quotas in Norway. Thus the increase in private governance across multiple arenas has been comparatively 'naturalized', meaning that influencing it has not been considered possible. At the same time regulation through democratic systems is increasingly criticized and targeted for potential political action.

6.1.4. Is globalization a key determinant of change?

The above sections have described changes as outcomes of economic and political globalization. One reason that such a finding is possible is of course the broad and also somewhat nebulous nature of globalization as a concept. Given that globalization refers to a process that alters the role of the state and increases the role of the market, it applies to most of the stakeholders' statements that relate to economic and political change in this direction. Change towards larger-scale markets, for instance, is congruent with the expected and actual impacts of globalization: the image of market relations and the role of the state vis-à-vis that of international-level systems would presumably have been very different, say, 50 years earlier.

This broad concurrence does not mean that all change can be attributed to globalization, however: limitations on the impact of globalization have been noted in the continued role of the state in determining much of resource use. Interviewees also described changes under economic and political change that could be indirect results of globalization but need not be interpreted as such: the degree to which they would depend on the decision-making process and whether decision-makers implemented the changes in response to internal administrative requirements or as restructuring in response to larger-scale globalizing processes (making it difficult to differentiate internal and external requirements). Among the changes described by stakeholders in this regard were that the degree of sectoral regulation has increased and greater administrative demands have been placed on local actors. This change mirrors an increased pace overall in resource use and decision-making. To cite an example, reindeer herders today need to coordinate their work with many other areas of land use, such as municipal planning, which requires them to be knowledgeable about several different aspects of land use, legislation and regulation. Entry into the occupations has also become more difficult in all the areas: the gate-keeping mechanism determining who gains entry into a particular sector/livelihood operates through regulation, for example fishing quotas, limits on the number of reindeer husbandry units (*driftsenheter*) in Norway and decreased recruitment, as limited profitability discourages new entrepreneurs from entering the sectors.

In general, there is concern among the sectors about the limitations on the use of resources, a restriction that is often enforced through the governance networks and relates at least partly to pressures to increase outtake. Sometimes, these pressures are a result of demands among multiple users, sometimes of increased market demands due to globalized competition; in one instance in the fishing sector, increased pressures can to some extent be seen as the result of increased natural competition with an invasive species. Such concerns over limited resources can be seen in each sector: forestry in Sweden is concerned with conservation limiting available forest resources; the fishing industry in Norway is worried that the king crab invasion and resultant environmental changes will limit fish resources; and reindeer herders are uneasy that overgrazing and competition from several sectors have reduced the amount of pasture land available for herding. All these factors highlight the interconnected nature of socio-ecological systems at the same time as they underscore some of the problems of attempting to manage resource use in a sector. In each case, interviewees were concerned at the practical level about all changes that decreased their access to resources, including managerial changes such as environmental protection that might in the longer term actually contribute to maintaining the resource.

An additional consequence of limited resource access is that increased competition and fewer but larger-scale actors has resulted in an 'acculturation' of nature, which is an adaptation to the limited resource situation overall, but one that is only indirectly attributable to globalization. Examples of this change are the movement towards reindeer farming in reindeer herding and, to some extent, that towards fish farming in fishing. In other words, both reindeer herding and fishing are moving towards a larger share of controlled populations and production through feeding in order to limit natural variability in production. However, these actions do not necessarily affect the market system, where increased supply may lead to lower prices or other market changes may similarly alter the context for local actors' adaptations. Moreover, more uniform and stationary reindeer and fish populations may be at greater risk from pests, for example. Adaptations that are economically relevant in the current situation may thus result in new actual or potential risks. Interestingly, despite the fact that the focal industries involve renewable resources, adaptations only rely on natural systems to a limited extent, and many actors have attempted to reduce or eliminate their reliance on systems with a high natural variability.

6.1.5. Is globalization a key determinant of social vulnerability and adaptive capacity?

The foregoing arguments show that social vulnerability – the 'baseline' vulnerability and adaptive capacity that populations exhibit today – is impacted to a considerable degree by globalization, although some factors (such as the market) are more apparent and influential than others. Through its impact on governance networks and the selection of actors to be included in decision-making in the sectors, globalization can also be shown to have an impact on the general determinants of adaptive capacity. The way in which the governance network is structured – to some extent as a result of globalizing forces – affects factors such as technology, material resources, political capital and entitlements, or resource rights. Table 6.1 lists the general determinants of adaptive capacity presented in Table 1.1. The table further indicates the factors identified in the present study that may be associated with each determinant and assesses the impact from economic and political globalization on each determinant in light of the present research.

Globalization could thus be seen as shifting the pattern of vulnerability, especially concerning economic factors impacted by market competition, and as shifting the pattern of governance with regard to actors and levels of decision-making capacity. This study has examined particular renewable resource-based sectors. While these could be expected to focus on their own interests as expressed in economic terms, they may also capture more broadly concerns of individual actors who are acting to maintain their present livelihoods within a market system. If the pattern observed here were also to hold true in other areas, it would mean that capacities, such as the ability to adapt either individually or structurally to international market changes and globalizing features, would be crucial for decreasing vulnerability; however, such a mechanism would be most relevant for areas or sectors dependent on economic livelihoods (with perhaps other mechanisms influencing subsistence livelihoods, which thus far have been a major focus in both Arctic studies and community vulnerability studies, for instance, Ford and Smit, 2004). This study thus adds an example of the different processes that may impact adaptive capacity in market-based as opposed to subsistence occupations.

Globalization can also be seen as a mechanism that links and explains the processes of change in otherwise generic and often unconnected determinants of adaptive capacity, at least on the scale of the governance network; however, it is less directly or immediately relevant for factors related to human capital, such as education levels, health or individual risk perception. While social capital is undoubtedly important for interaction within governance networks, this study has shown that the existence of local networks (through which, for instance, refinement or a united local perspective for local resource rights could be developed) may also be limited by competition both within and between sectors. It may be that local social networks linked to an occupation have a more limited role in providing support for the individual actor where, as here, adaptations

Table 6.1 *Impacts of globalization on determinants of adaptive capacity*

Determinant	Encompasses (examples from the local case studies)	Impacts from globalization
Human capital	Traditional knowledge in the sectors becoming less important Education level and other capacities that enable integration into wider networks and understanding of the regulative and administrative system	Limited direct impact (possible indirect impacts from effects on industry)
Information and technology	Industrial and applied technology that is utilized in the sectors (determining the form and level of rationalization as well as the increase in efficiency possible for individuals due to technological access) Innovation capacity/entrepreneurship determines the possibilities for creating diversification and spreading into new niches of employment	Major impact in structuring patterns of competition and technology requirements (e.g. supporting increased efficiency)
Material resources and infrastructure	Underlying capacity for production and the transport of production to markets Local infrastructure, encompassing the broader employment structure, tax base and population characteristics	Major impact from effects on employment structure and capacity for production
Organization and social capital	Local coping networks limited by competition and, in some cases, focus on individual enterprises	Impact on state–civil society relation and social mobilization
Political capital	Limited density of relationships and decision-making capacity	Impact on political networks as a part of governance and on shifts in these
Wealth and financial capital	Limited access to financial instruments, although some government budgetary transfers exist	Impact on economic structure of competition that determines availability of financial instruments
Institutions and entitlements	Property and resource rights crucial	Impacts on informal and formal rules for resource use, e.g. through norms

Source: The main determinants of adaptive capacity are adapted from Eakin and Lemos (2006), p10, and Naess et al (2006), p224

are largely targeted at market-based competitiveness for individuals or companies; actors draw extensively upon technological means for adaptation, and the welfare state is able to provide for basic security that limits immediate reliance on the local community (cf. Pelling et al, 2007). This underscores the differences in determinants of adaptive capacity cited by Naess et al (2006) for an advanced industrial economy (among which employment figured prominently) to generic determinants of adaptive capacity. In order to identify the processes that determine adaptive capacity, including globalization, it may thus be crucial to draw a distinction between subsistence and market-based economies, as globalization may have a much more marked influence on the latter.

6.1.6. Climate change

As a specific impact on the case study sectors and areas, climate change and its consequences are discussed differently by different actors, with views largely varying by sector and the degree of detachment or closeness of the individual interviewee to the natural environment. These findings underscore the results in adaptation research that adaptation is not undertaken in response to climate change in isolation but in response to the actors' perceived situation as a whole. In this work, the ways in which actors adapt have been shown to be continuous and to mainly comprise responses to factors impacting their economic livelihood. Adaptation directly to climate change is also mediated (sometimes heavily) by governance systems that set the parameters for resource use, and the direct impact of climate change – even on renewable resource use – is sometimes decreased by a decoupling of the production system from nature. An example of both processes is provided by fishing in northern Norway, where fishermen suggested that, owing to small fish quotas, fluctuations in fish populations due to climate change may not fundamentally have an impact on the individual fisherman. Increased fish farming signals a further decoupling from the natural system. A similar development can be seen in reindeer herding, where increased farming or additional feeding is limiting reindeer herding's reliance on the natural environment. Additionally, people's focus on more long-term changes was limited compared to the more urgent demands in the sectors (primarily the direct economic ones).

A wide range of environmental changes and risks was noted among those who work in the field, such as forestry machine drivers, reindeer herders and fishermen. The specific impacts of potential climate change seemed to have the greatest effect on perceived vulnerability among reindeer herders, as reindeer are affected even by day-to-day weather conditions and weather and the related grazing opportunities are a major determinant of production. Interviewees working in practical duties in forestry and reindeer herding also described somewhat similar vulnerable periods in autumn and spring that result from thawing and freezing. Freeze–thaw threshold conditions were also generally among those described as affecting the areas most, as they involve the transition between water and ice or snow, which has implications for the entire environment. Interviewees working in forestry and fishing, however, even though those sectors could potentially be strongly impacted by projected climate changes, expressed more limited concern about potential climatic changes than did those in reindeer herding. This more limited concern could be attributed to factors such as large variability and limited resource outtake (in fishing), limited awareness and planning for longer-term adaptation, and the strong impact of socioeconomic and political factors on the two sectors.

Forestry stakeholders, for instance, emphasized changes affecting road and ground accessibility. Interviewees in the fishing industry stressed the variability of the sector, which makes trends difficult to perceive and requires constant adaptation to changes. Interviewees in fishing also suggested that for larger changes, such as changes in fish species and their distribution, adaptations need to be managed principally through the regulation system (by, for instance, releasing quotas for new species to northern regions) and, to the extent manageable, through individual economic adaptations.

The results suggest that climate changes in themselves or adaptations to them are not highlighted as such if they are not seen as impacting production. Climate change impacts were emphasized and adapted to primarily when they (alone or in connection with other factors) affected stakeholders' economic situation (quality and quantity of production). Given this linkage to the wide variety of resources possessed by an actor, the actors with the most limited economic and political capabilities are also those who are most vulnerable to climate change impacts. Economic, political and climatic changes thus largely affect similar actors: small-scale actors with the least resources for adaptation are losers in processes of change; the winners seem to be large-scale actors (O'Brien and Leichenko, 2000). Large-scale actors are not only better resourced economically but also able to draw upon larger interest groups and political networks to gain or maintain rights to resource use, and can also afford the reorganization costs that may come with adaptations or specific actions to increase their competitiveness. For instance, large-scale actors in forestry may be able to choose their access roads from a large road network and have the resources for improving road structure and sanding, thus gaining enhanced increased opportunities for harvesting throughout the year. In the case of reindeer herding, if climatic changes increase fluctuations across the freeze–thaw threshold, actors who can afford supplementary feeding (due to lower fodder costs or larger monetary resources to purchase fodder) or who have reserve grazing areas may be better off.

Additionally, the results for the three sectors suggest that a number of multi-faceted interactions between different types of stimuli and adaptations make the value of any one adaptation difficult to assess. One such instance is the interaction between socioeconomic, political and environmental/climate factors observed in the case of reindeer herding: as improved weather conditions for reindeer herding have led to increased meat production – supposedly a benefit – the market price has fallen, resulting in a crisis. This shows not only that many of the focal adaptations are individual economic adaptations, in other words ones that actors can carry out themselves, but also how difficult it is to assess the outcome of adaptations beforehand. Overall, as possibilities for open – or at least less controlled – access to resources are diminishing, climate change that affects the use of resources in clearly demarcated areas may have substantial impacts. Any impacts of climate change on the use of resources will become amplified where they occur in clearly demarcated areas. Another development that will affect a range of actors is that climate change will very likely exacerbate conflicts in the economic and political sectors as the access to resources decreases. For example, climate changes could well aggravate conflicts between forestry and reindeer herding if forestry finds it necessary to log more readily accessible areas that could otherwise have been left uncut for grazing or use as migration routes. Such changes could potentially impact public opinion and political decisions, possibly shifting patterns of governance and resource access.

Adaptations may thus not be readily distinguishable for each scientifically categorized climate change impact or be attributable directly to climate change (Smit and Wandel, 2006). They take place in response to the situation as a whole; different elements can of course be distinguished, but the climate has thus far been shown to be of limited importance in actors' minds (with the possible exception of those engaged in reindeer herding who use limited or no supplementary feeding). Instead of adapting to distinct factors, then, people may be seen as adapting continuously, with a focus on what they need to do to maintain or improve their current livelihood and socioeconomic situation.

6.2. Conclusion: Summary of vulnerability, adaptive capacity and types of adaptation in the areas

All in all, adaptations were essentially devised by the individual actors as an integrated part of retaining competitiveness in the economic market in which their livelihood was integrated and, ultimately, with a focus on maintaining that livelihood. The different adaptations and limitations on adaptation may be seen as indicating an 'adaptation space' (Berkhout et al, 2006) or adaptive capacity, as well as the changes that could be undertaken (often within the regulative system) to increase this space. The resources for adaptation were primarily monetary and technological (such as those applied in order to improve efficiency) but also included opportunities to influence and utilize regulation and the relevant governance structures and institutions, some of these at the international level. The interviewees also exhibited a very clear knowledge of what their problems were and how these could be alleviated but often found that the available monetary means or capacities afforded through regulative or broader governance systems fell short of their needs, at least for longer-term adaptation. In many of these cases, this means that the limit on efficient adaptation is often set by larger systems, and that efficient adaptation (as seen by local actors) cannot be undertaken without access to decision-makers – which some try to achieve through linkage with interest groups – and methods for inter-scalar integration and communication.

On a general level, the individual economic adaptations considered by the stakeholders can be summarized as follows:

- *diversification* (individual and societal);
- *concentration* on the most profitable practices: a trend contrary to individual diversification that aims to raise profitability;
- *introduction of additional technology* for more efficient production;
- *increased marketing and market orientation*: targeting the market rather than the production system in order to obtain higher prices per unit through customization; and
- *reduced dependence on natural conditions*: increased decoupling or detachment of production from natural conditions, for example supplementary feeding of reindeer or fish farming.

Economic adaptations thus occur for the most part at the individual level, the level at which interviewees considered it possible to undertake adaptations. To the extent that larger-scale or societal adaptations are mentioned, these are often linked to the framework or regulative and political parts of governance (often at the state level). Such adaptations would increase actors' possibilities for economic self-sufficiency, with the increased economic leeway lessening their dependence on the political decision-making network. Here, local actors are constrained by the limited support network available to them, the limited opportunities to influence the decisions that affect them and their low degree of interconnectedness; indeed, they suggested that, in the main, local adaptive capacity could only be significantly strengthened if such changes were implemented by actors elsewhere.

The larger-scale adaptations considered by stakeholders can be summarized as follows:

- attempts to increase *resource access*, for example changing legislation or other aspects of the regulatory framework;
- *coordination* with other actors, for example to reduce conflicts and gain boundary and coordination benefits;
- *maintenance and improvement of municipal services*, for example in order to attract new people to the areas;
- *societal differentiation in employment*, for example decentralization of state services and their relocation in the focal area;
- *state subsidies and other support*, for example the financing of investment through state loan institutions;
- *increased local/regional organization*: the development of organization and support networks for increased influence that allows for control over external actors such as environmental protection interests; this would include the development of a leadership culture; and
- *national-level changes in regulation and the development of local or regional resource management*: the decentralization of management to lower levels of government or cooperation bodies, in keeping with the proximity principle.

Larger-scale adaptations that interviewees would undertake to increase local adaptive capacity largely involve the state and state responsibility for the local area, as well as – albeit to a lesser extent – the need for increased coordination locally. As a result, local adaptive capacity is limited and requires adaptation at other levels to deal with many of the stresses that manifest themselves locally.

While individual economic and larger-scale adaptations are largely seen as interlinked (for example individual economic adaptations undertaken within the political framework that sets some limits on profitability), direct environmental adaptations are mainly undertaken to the degree that they will improve the economic situation of the actor. Adaptations are thus conceived of and invoked in response to the situation as a whole. Accordingly, general-level adaptations to climatic impacts largely relate to economic and political adaptations and consist predominantly of the following:

- *diversification and finding new customers/markets for new types or grades of products*;
- *reduced dependence on the local natural system*, for example the use of additional feeding; and
- *shifts in timing and location of practices.*

In the case of climate change, the principal constraints are costs and limited awareness of and planning for longer-term risks.

On the whole, vulnerable actors are those who are economically the smallest, have the most limited networks with restricted access to decision-makers, live in communities with limited scale benefits, have the most limited monetary resources, and have the most limited means to find additional resources, for instance through diversification. Diversification is not used here only in the sense of individual small-scale diversification into several occupations while practising one's livelihood proper; it also refers to socioeconomic diversification in the community at large, which would enable people to take up different, full-time employment, as this may be a relevant alternative for some of the sectors in an industrialized economy. The most vulnerable actors are also negatively impacted by globalization, as globalization benefits those who can work within international market systems, and operate out of larger localities and in more extensive networks in order to attain scalar benefits.

Overall, those actors can be considered especially vulnerable to change whose overall vulnerability across the range of determinants is high and who cannot, as other individuals have done, compensate for remote location and limited local resource access by tapping an extensive international buyer network that yields added value. While individual capacities may thus compensate for a marginalized location, generalized impacts on a location from globalization show a strong effect. Here, globalization can be seen as explaining some of the pattern of impacts on the general determinants of vulnerability. Vulnerability at the level of individual actors is to a large extent determined by, among other things, governance factors and geographical location within a governance network (including access to markets and scale benefits for the type of community in which the actor operates). A lack of such alternatives may limit people's possibilities to find new employment in the locality where they live, and may prompt them to migrate although their preference would be to stay in the locality.

However, this highly differentiated vulnerability among different actors in the same localities poses the question 'vulnerability for whom?'. From the perspective of his or her livelihood, a reindeer herder may be highly vulnerable if he or she cannot access resources or arrange a buffer to survive difficult periods, for example unfavourable weather conditions, and faces the prospect of going out of business as a result. On the other hand, the closure of a large-scale industry may entail a larger vulnerability at the regional or local scale, as it would affect a larger part of population in an economy with limited options for diversification, possibly further increasing vulnerability as a result. These arguments illustrate that it is only in the context of a broader social vulnerability and systemic processes that one can distinguish beneficial adaptations from detrimental ones, even for particular actors. This also holds true for the context in which adaptations within a sector have to be assessed. For example, the adaptation to variable environmental conditions through supplementary feeding in the case of reindeer herding could be seen as an effective one from the viewpoint of climate change impacts; however, when viewed in the context of the market system and its reaction to a surplus of reindeer meat, supplementary feeding can, in some instances, be seen as a costly adaptation that is not fully compensated for by the market. An additional question regarding 'vulnerability for whom?' is whether processes to counteract the limitations to adaptation that have been defined here would increase or limit adaptive capacity for other actors and interests. Such an assessment would be crucial in order not only to advance the interests of particular sectors, groups or regions, but also to accomplish trade-offs in policy. This in particular speaks to the political nature of vulnerability, where who is vulnerable is necessarily determined (and assessed) in a political arena of conflict within and between groups. It also speaks to the fact that an increase in adaptive capacity among certain actors or interests does not necessarily reduce environmental vulnerability, as the adaptive capacity of specific actors may be raised at the cost of a higher utilization or consumption of environmental resources, for instance, wood or other forest resources.

For small-scale actors in a locality, the capacity to organize locally to counteract larger-scale interests or processes, or to develop a local entrepreneurship or leadership culture, has been mentioned above as a way of increasing adaptive capacity and possibly also an actor's opportunity to influence higher levels of decision-making (for example in the state). This capacity may, however, be weakened by the degree to which resource conflicts cause fragmentation or divisiveness among actors locally. The existence of resource conflicts in general tends to spark discussions of rights and how resources should be distributed, including how extensive the rights should be and who is entitled to what. This observation makes it relevant to look at how conflicts play out locally: do they – for example in the case of international norms – aggravate ongoing conflicts, which may politicize issues and make cooperation even more difficult? The interviews conducted for the present research indicate that groups at the local level become divided along the lines of resource access if these are drawn locally; otherwise, the local level may organize against what are perceived as larger-scale, non-local interests. There are thus no absolute borders and settled definitions which would suggest, for instance, that ethnicity as such is the source of conflict; rather, conflicts are inherently bound up with the policies on and requirements for resource access or entitlement. Ethnic conflict in the areas may thus be seen as related to the application of and distinctions embodied in policies on resource access rather than to ethnicity or indigenousness as such.

The study undertaken here has attempted to bring to light some of the complexity of vulnerability and the frameworks with which one must interact to squarely address the complexities involved and to give due consideration to the multi-level character and real political and economical interests of the stakeholders interviewed. The insights gained through this approach include an understanding of the available adaptations seen by stakeholders, the way in which these could be extended by actions on other levels, and the scope for adaptations that exists or could be opened through actions on larger scales than that of the locality. One salient realization prompted by this study for participation and policy advice processes is that stakeholders cannot be assumed to interact on their own volition at the local level, across levels and in their regular social context in ways that are not already institutionalized or not considered readily accessible by the stakeholders themselves. Additionally, the research findings add to our understanding of how natural resource utilization systems may be decoupled from natural conditions or governed by regulation (such as quota systems) to such an extent that climate change impacts may not immediately affect resource users and responses to climate change may be severely delayed. The study has also illustrated the practical limitations on access to decision-makers, the crucial importance of governance structure and external impacts such as globalization, and the potential differences in the processes determining adaptive capacity in subsistence versus market-based occupations. The research has also brought to light the large number of institutions on different levels that act on or constrain resource access, adaptation and agency within the sectors studied and thereby fundamentally limit the possibilities for adaptation that exist at the local level. Ultimately, one application of the findings here could be to problematize organization and interaction among groups as well as the dynamics determining adaptive capacity.

Notes

Introduction

1 Adaptation, which, like vulnerability, is a central theme of the second working group of the IPCC (IPCC, 2001), has received relatively little attention in empirical studies in comparison with mitigation (Jepma and Munasinghe, 1998; Tol et al, 1998; Fankhauser et al, 1999). Mitigation – ways to hinder climate change – has gained the considerable attention given to it in the literature by most often being connected to decision-making at a national or global scale and to large-scale cutbacks in emissions, which can be dealt with at the global and national political levels. But while mitigation is also possible at the local and regional levels through changes in lifestyle, such as decreased car use, adaptation may be the crucial mechanism to look at on these levels. Another reason why adaptation is a crucial focus is that the local and regional context is ultimately the one in which impacts will become visible even if emissions are stopped immediately (IPCC, 2001).

2 As Turner et al note, 'The use of place-based approaches makes obvious the need to find methods to operationalize vulnerability analysis that are useful for the specificity of place and for building general concepts from them' (Turner et al, 2003, p8076).

Chapter 1

1 For instance, to date, 'there [has been] little research on the dynamics of adaptation in human systems, the processes of adaptation decision-making, the conditions that stimulate or constrain adaptation, and the role of non-climatic factors' (Smit and Pilifosova, 2001, p905). 'There also has been little research to date on the roles and responsibilities of individuals, communities, corporations, private and public institutions, governments, and international organizations in adaptation' (Smit and Pilifosova, 2001, p905; cf. Tol et al, 1998). Similarly, 'while many types of adaptation have been distinguished, there is less scholarship on actual adaptation processes' (Smit et al, 2000, p241).

2 The concept of path-dependency, used in several disciplines, emphasizes the persistence and tenacity of institutional arrangements and has some relation – although

little developed – to adaptation. Path-dependency centres on the tenet that once a particular institutional set-up and traditions have been developed for a particular practice, it is difficult to change tack (Löfgren and Benner, 2003); existing paths thus determine and delimit future choice to a certain extent.

3 Similarly, Adger et al note: 'We must ensure that countries are able to cope with existing hazards and those anticipated in the near term, in order that damage from such hazards does not hold back development efforts and exacerbate existing vulnerability, undermining the foundation on which adaptation to future climate change must be based ... Reducing vulnerability to existing hazards is therefore the most desirable starting point for reducing the risks associated with climate change' (Adger et al, 2004b, p42).

4 Similarly, Osman-Elasha and Sanjak note: 'The social and individual capacity that is being built for community action, flexible management of natural resources and diversification of livelihoods will be valuable for years to come, whatever the climate may be' (Osman-Elasha and Sanjak, 2007, p255).

5 Definitions of livelihood sustainability sometimes approximate sustainability definitions on an individual scale. For instance, 'livelihood is sustainable when one can cope with and recover from stresses and shocks and maintain or enhance one's capabilities and assets both now and in the future, without undermining the natural resource base' (Osman-Elasha and Sanjak, 2007, p245).

6 Liverman notes this rationale succinctly: 'To be vulnerable to drought is to lack environmental, technological, economic or physical defences against its impact' (Liverman, 1990, p50).

7 For instance Moench notes that: 'If, as our research in South Asia suggests, adaptive capacity and livelihood resilience depend on social capital at the household level (in other words education and other factors that enable individuals to function within a wider economy), the presence or absence of local enabling institutions (local cooperatives, banks, self-help groups), and the larger physical and social infrastructure that enables goods, information, services and people to "flow", then interventions to catalyse effective adaptation are important at multiple levels' (Moench, 2004, p13).

8 In a similar vein, Stonich and Bort note for a Central American case that 'larger operations predominate. Regardless of country, a similar set of constraints affects the success of small farmers: government concession, pricing and subsidy policies; inadequate access to capital, credit, technology and technical assistance; and unfavorable insertion into the market' (Stonich and Bort, 1997, p169).

9 Smit and Pilifosova note: 'The main features of communities or regions that seem to determine their adaptive capacity [are] economic wealth, technology, information and skills, infrastructure [including availability of and access to resources by decision-makers], institutions, and equity' (Smit and Pilifosova, 2001, p895).

10 Similar factors have been discussed by other authors, for example Moench and Dixit, 2004; cf. Moench, 2004; Cutter et al, 2003.

11 An example of this is early work on 'dumb farmer–smart farmer' assumptions in the case of adaptation (cf. Smit and Pilifosova, 2001).

12 For instance, given the major complexity especially where this social dimension is concerned, some have noted that 'to explain human action as adaptation is commonly to under-specify the social and economic forces at work in changing behaviour' (Thompson, 2006, p2).

13 It would be possible to describe what is here called 'globalization' as 'internationalization', as very few of its manifestations are truly global but, rather, entail increased interactions within limited areas, such as a continent or international region. Some argue, however, that globalization entails 'a functional integration of internationally dispersed activities' that is qualitatively different from mere internationalization, involving 'the simple extension of economic activities across national boundaries' (Dicken, 1998, p5, in Rankin, 2003, p710). Globalization in this sense would include greater interconnectedness than would internationalization.

14 As Roudometof notes: '"Globalization" is a term currently used in an unstructured manner to indicate a series of economic and policymaking trends characteristic of the 1990s ... "globalization" is often used as a present-day equivalent of "modernization" ... there is no clear-cut distinction between the prescriptive strategies advocated by self-proclaimed "globalization" theorists and post-World War II "modernization theory"' (Roudometof, 2003, p38). A distinction can, however, be drawn between the cross-border, integrating globalization of the late 20th century and the previous expansion of the world economy that took place mainly between mutually exclusive, geographical national markets (Guillén, 2001). The concept of modernization will not be dealt with further in this work except where it is germane to note that there exist different understandings of the connection between globalization and modernity.

15 In contrast to economic globalization, political globalization may be seen more as a regional internationalization than as globalization: many of the organizations and standards formed are of limited rather than universal reach (Budd, 1998).

Chapter 2

1 This work takes an interpretative or 'understanding' approach, which differs from 'explaining' approaches. Hollis and Smith define the difference between 'understanding' and 'explaining' in another field of study thus: 'One story is an outsider's, told in the manner of a natural scientist seeking to explain the workings of nature ... The other is an insider's, told so as to make us understand what the events mean, in a sense distinct from any meaning found in unearthing the laws of nature. ... "Explaining" is the key term in one approach, "understanding" in the other' (Hollis and Smith, 1991, p1).

2 Some view traditional ecological knowledge as indigenous, some as both indigenous and local, still others, such as Tengö and Belfrage, note that 'ecological knowledge used in local resource management can also be generated by and reside in communities that lack historical and cultural continuity' (Tengö and Belfrage, 2004). The present study potentially deals with all of these different types of knowledge and will not apply an explicit differentiation.

3 Open-ended interview questions have sometimes been described as representing an ethnographic approach, suitable for science-related perception research in particular. For instance, Kempton notes: 'Ethnographic methods were developed by anthropologists to study foreign cultures, but I have found in my previous work that an equally important application is in bridging the gap between the lay person and the scientist in understanding science policy issues' (Kempton, 1991, p184).

4 In some instances interviewee selection was also based on availability. Stora Enso, for instance, the relevant large forest company to interview in northern Finland, refused

to participate in the study (possibly following media attention around the time the interviews were conducted).

5 A majority of the interviewees were male, as the focal sectors are male-dominated. Particular efforts were made to include female respondents in the study, however. Interviewees were also selected based on the duration of their employment or work in each sector, so that they would be able to elucidate change over time. Interviewees had typically worked for about a generation in the sector (all of their working lives) and were typically middle-aged or older. Each interview lasted from one to one-and-a-half hours.

6 The reason for adopting this practice was that some of the localities where interviews were conducted are so small that an indication of organization and location would in effect make it possible to identify the respondent. Interviewees in smaller localities will thus be referred to mainly by region.

7 Schiller et al provide a salient account of lay-language indicators that exemplifies the importance of communication skills and the gap between scientific and lay language and characterizations that may impede discussion. They note that ecological assessments and monitoring programmes often rely on indicators to evaluate environmental conditions: 'Such indicators are frequently developed by scientists, expressed in technical language and target aspects of the environment that scientists consider useful. Yet ... making environmental decisions requires ... consideration of what members of the public value about ecosystems' (Schiller et al, 2001, para 1). They further observe, based on a series of small-group sessions, that people did not want to know what these indicators measured or how measurements were performed. Rather, respondents wanted to know what such measurements could tell them about environmental conditions. Most positively received were descriptions that included various combinations of indicators and gave a user-centred illustration of ecological conditions on the whole. Thus, for instance, the lay term 'timber production' was used by Schiller et al to cover a number of scientific indicators, including soil classification and physiochemistry, leaf area, canopy diversity, crown measures, branch evaluations, lichen chemistry, foliar chemistry, regeneration, dendrochemistry, mortality, and visible damage (Schiller et al, 2001). Schiller et al conclude: 'As an example, our investigation suggests that non-scientists may find information on the changing levels of "contamination of forest plants by air pollution" in a region to be more salient than specific information on the individual measures used by EMAP that together inform that topic (for example foliar chemistry, lichen chemistry, dendrochemistry and branch evaluations) (Schiller et al, 2001, paragraph 51).

8 The local papers included are *Piteå-Tidningen* in Sweden and *Lapin Kansa* in Finland, which are the largest local papers in the respective areas. These were selected to allow for a comparison with the interviews in regard to questions dealing with forestry and nature reserves and to indicate whether these issues were discussed more broadly in the study areas in a way comparable to that in which interviewees treated them in their statements. In Norway, the newspaper *Ságat*, which is aimed at the Saami population, was included. This paper was selected to provide for a comparison with interviewees' statements on indigenous issues in the areas and to indicate whether these views were more broadly representative of indigenous views in the area.

9 In this regard the study also shows some similarity in approach to Marsden's work, which assessed how environmental vulnerability in the agricultural sector in a

Caribbean case study was being created through particular socioeconomic adaptations. Marsden's study focused on qualitative interviews with key participants in food production and supply networks: 'The data provide a basis for assessing the activities of key actors in the food networks, focusing most especially on their coping strategies and on their relationships with other actors in different locations in the networks both locally and at a distance' (Marsden, 1997, p324).

10 The work will not explicitly use the social network approach applied by Ziervogel and Downing (2004), which was chosen for a case dealing with innovation diffusion. The aim here is to map the type of actors (business, administration or interest organization) and the organizational scale on which they operate (local, regional, national or international) as seen from a local/regional perspective, for which other means of illustrating linkages have been chosen.

Chapter 3

1 In Sweden, the state organization Fastighetsverket owns and manages protected areas, but this organization and its role were not discussed by interviewees, whose focus was productive forest.

2 At the time of the study, the organization in Sweden consisted of ten county forestry boards and their districts, falling under the National Board of Forestry. As of the start of 2006, this organization changed and now, among other things, encompasses larger district boards than previously, organized under the Swedish Forest Agency (Swedish Forest Agency, 2006).

3 Stora Enso declined to participate in the study, possibly as a result of some criticism towards the company in the media at the time the interviews were conducted.

4 Original text: 'Täällä Itälapin alueella on muistaakseni 99 metsuria työllistetty 5 kunnan alueella' (Metsähallitus, planning).

5 Original text: 'suurin muutos oli kun uittoa lopetettiin. Se oli yksi suuri muutos koneistamisen ohella' (Common Forest I, Finland).

6 Original text: 'när man började sjuttiosju så kunde jag aldrig tänka sig att en maskin skulle avverka' (Sveaskog, Sweden).

7 Original text: 'näitä hydraulinostureita niin vanhat automiehet sanoivat että tällä ei tee mitään ja peruste oli se että se tekee niin rumaa jälkeä' (Transport Entrepreneur, Finland).

8 Original text: 'Nythän yksi mies tekee hirveän määrän sillä nosturilla töitä päivässä. Siihen pitäisi olla sadan miehen lauma sitä vastaan sitten' (Transport Entrepreneur, Finland).

9 In original: 'kädestä suuhun' (Transport Entrepreneur, Finland).

10 Original text: 'todella suuria ja automatisoituja sahoja' (Local Government, Finland).

11 Original text: 'de här ... mindre inlandssågarna, de har ju ingen chans att konkurrera' (County Forestry Board, Sweden).

12 Original text: 'likviditetsbrist ... man hamnar i någon slags situation där du inte kan du har dålig lönsamhet och ska försöka bli större och så ökar du volymen och så blir timret för dyrt och då får han ännu sämre lönsamhet och banken liksom vill dra fingrarna ur syltburken så fort som möjligt' (Sawmill Owner, Sweden).

13 Original text: 'Virossa ja Etelä-Suomessa saa haapaa viljeltyä kanssa noin 25 vuodessa' (Common Forest II, Finland).
14 Original text: 'ei ole kannattavaa eikä kannata kasvattaa koivua täällä hitaasti' (Common Forest II, Finland).
15 Original text: 'massakapitalindustrin bara i Piteå ligger på en åtta miljarder och det är ju ingen självklarhet att de är kvar där, utan de kan flytta där råvaran finns' (County Forestry Board, Sweden).
16 Original text: 'de tar ju från baltvirke idag ... [och] ungefär en miljon kubikmeter får vi från västerbottensskogen ... visst är det en känslig fråga inte minst för politikerna här' (County Forestry Board, Sweden).
17 Original text: 'bristen på talltimmer idag den är ju 1,1 miljon kubikmeter ... och om den då tillförs ifrån Ryssland eller Finland eller Västerbotten eller vad du har för nåt det är egalt' (Älvsbyhus, Sweden).
18 Original text: 'det är mycket litet som vi har tagit från Finland därför att det är för hård konkurrens [med] finska sågverksägare' (Älvsbyhus, Sweden).
19 Original text: 'globala inom företaget' (SCA, Sweden).
20 Original text: 'så det är klart att den svenska marknaden är jätteliten för oss' (SCA, Sweden).
21 Original text: 'om det blir alltför tufft och kämpigt så finns det ju andra ställen, så varför ska man stånga sig blodig här' (SCA, Sweden).
22 Original text: 'Älvsbyhusandan' (Älvsbyhus, Sweden).
23 Original text: 'se olisi ihan katastrofi jos se tehdas loppuu' (Lapland Forest Centre, Finland).
24 Original text: 'Onhan se selvää että sieltä voidaan tuottaa puuta halvemmalla jos ei ole uudistamisvelvoitteita samalla tasolla mitä meillä, ajattelin esimerkiksi Venäjän valtiossa' (Forest Owners Union in Northern Finland).
25 Original text: 'tietenkin metsävarat on Venäjällä kovat, mutta jos teollisuus siellä nousee ja noin päin pois, niin kyllä kai ne itsekin sitä tarttee ... sieltä on puutavara hankittu noille Itä- Suomen tehtaille, että tietenkin ilman sitä puuta sieltä niin eihän se tietenkään näin iso olisikaan ilman ulkomaista puuta' (Common Forest II, Finland; cf. Common Forest I, Finland).
26 Original text: 'Savukosken kunnalta puuttuu sellainen yrityskulttuuri kun esimerkiksi Pohjanmaalla jossa joka talossa on yritystoiminta. Yrityskulttuuri puuttuu ja mentaliteetti on erilainen' (Metsähallitus, planning).
27 Original text: 'de som tar besluten kan sin grej, han som tar beslutet om investeringar har avtal med kunder för sönderdelad produkt ... Man sitter med sitt och tänker på sitt' (Local Government II, Sweden).
28 Original text: 'aluetaloudellisesti, tämä toiminta, tämä on hirvittävän tärkeä koska se tuottaa, mutta ne työpaikat on vain jossakin muualla sitten' (Metsähallitus, planning).
29 Original text: 'ledningarna för ... Sveaskog har ju varit upp och tittat men det är ju så att när jag säger att ni äger sjuttio procent av kommunen och har ett visst ansvar så håller de ju med mig men däremot att se några åtgärder just efter att de har det har ju aldrig varit ... de säger ju just det att det är ju inte vi utan det är de egna [dotter]företagen sen säljer de ju ut företagen' (Local Government II, Sweden).
30 Original text: 'Vaikuttaminen on vaikea' (Lapland Regional Council, Finland).

31 Original text: 'vår egen kraft är för liten och det är ju så att det här med avreglering av verksamhet det gör ju att där marknaden inte finns där finns inget intresse heller' (Local Government II, Sweden).

32 Original text: 'En byggnad avskriver du ju kanske på trettio år så vet du så någonstans finns ju det perspektivet då när du börjar eller ja vissa byggnader kanske femtio år' (Local Government II, Sweden).

33 Original text: 'vi har vår roll att se så att lagarna efterlevs ... lagarna är väl inte alltid skrivna som skogsbruket eller rennäringen tycker de skulle vara skrivna ... ibland har de stora förväntningar på oss som vi [inte kan uppfylla]... men vår absolut största uppdragsgivare det är samhället det är riksdag och regering' (County Forestry Board, Sweden).

34 Original text: 'vi har ju nått smärtgränsen' (County Forestry Board, Sweden).

35 Original text: 'skogsvårdslagen är ju inget problem alltså den är ju egentligen ganska tam jämfört med alla andra krav' (Common Forest, Sweden).

36 Original text: 'då har vi tappat i alla fall åtta skogsår' (Älvsbyhus, Sweden; cf. Common Forest II, Finland, Common Forest I, Finland).

37 Original text: 'De sa ju det att vi ska utnyttja framför allt råvaran till oss själva till våra egna enheter ... vi fick ... stå ... i dörrn, fråga om vi liksom kunde få köpa någon råvara' (Sawmill Owner, Sweden).

38 Original text: 'vi [är fortfarande]... utlämnade åt Sveaskog och deras policies ... De sågverk som är kvar ... de har alltså råvaruproblem ständigt ... och det har ju gjort att då antalet sysselsatta inom skogsnäringen har ju sjunkit' (Local Government II, Sweden).

39 Original text: 'förut var det mer att man funderade på hur man skulle kunna ta tillvara på skogen, det var produktion mer än reservat' (Sawmill Owner, Sweden).

40 Original text: 'nämä metsänluonnonhoitoon liittyvät ympäristöasiat josta ei puhuttu ennen kun 80-luvun puolienväliin päästiin' (Lapland Forest Centre, Finland; cf. County Forestry Board, Sweden).

41 Original text: 'det har blivit mycket mer intensivt måste man säga sedan nittio ... det har blivit en fullständigt ny indelning av vårt skogsbruk i grund och botten, ny indelning och då började man med att registrera naturvärden ... och sen en omfattande utbildning, vi anställde ju en egen ekolog då också ... det är ju stor skillnad ... sen är det ju det att vi skulle certifieras' (SCA, Sweden).

42 Original text: 'jos se on Natura aluetta niin se päätös hoidetaan muualla' (Lapland Regional Council, Finland).

43 Original text: 'just nu avsätts det väldigt mycket mer [skog] för att bilda reservat för naturhänsynen ... jag tror att just nu känner sig skogsbruket väldigt pressat det här med Natura 2000' (County Administration, Sweden; cf. SCA, Sweden)

44 Original text: 'så att vi skulle stå för en tredjedel av det här landets besparing och då förstår man att då undrar skogsindustrin att vart ska det här löpa' (County Forestry Board, Sweden).

45 Original text: 'heillä on ollut suojelun kanss ongelmia ja vaihtomaista saatu korvaus on riittämätön' (Common Forest I, Finland).

46 Original text: 'tämä yltiöpäinen suojelu, joka ei onneksi ole vielä kouraissut ... yhteismetsän kohdalla' (Common Forest I, Finland).

47 Original text: 'det drabbar ju lokalbefolkningen' (Forestry Machine Driver, Sweden; cf. Local Government II, Sweden; Metsähallitus, planning).

48 Original text: 'Sen takia meidän kuntamme on kauheassa pinteessä tässä nyt ollut koska tuo tulopohja on taloudellisesti muuttunut' (Local Government, Finland).

49 Original text: 'varje kubikmeter skog som avverkas har ett BNP-värde på ungefär två och ett halvt tusen, så att att det får konsekvenser, ekonomiska konsekvenser, om man så att säga drar bort skogsmarken från produktion' (County Forestry Board, Sweden).

50 Original text: '[J]ag är förste till att hålla med om att man ska ta miljöhänsyn men det får inte hela tiden bli på bekostnad av, man ska ha utökning av tillväxten för ... stryper man tillflödet då får man ingen tillflyttnad och tillväxt' (Sawmill Owner, Sweden; cf. Metsähallitus, planning; SCA, Sweden).

51 Original text: 'Om de börjar göra alla reservat som de har tänkt i det här området, då är det som att sitta i sitt eget skafferi och ändå inte få äta ... det ska man vara väl medveten om ska du bedriva industriell verksamhet i inlandet i sågverksbranschen då måste vi ha timmer. Annars ... tar vi [som företagare] ut pengarna och sitter och vickar tårna, men det finns ju ingen sysselsättning' (Sawmill Owner, Sweden).

52 Original text: 'Stora Enso jossa tämä konttori on Lontoossa ... sehän helposti ajattelee että hyvänen aikaa Lappihan on eurooppalaisten virkistysaluetta että ei kai sieltä saa nyt puuta kaataa hakatakkaan, että tavallaan tämmöinen' (Lapland Forest Centre, Finland).

53 Original text: 'den här sociala dimensionen' (County Forestry Board, Sweden).

54 Original text: 'bortom den traditionella skogsvårdslagen' (County Forestry Board, Sweden).

55 Original text: 'täytyy täällä hyväksyä tämmöinen monitavoitteinen yritystoiminta, ei niin että vain maataloutta, vain porotaloutta, vain matkailuyritystoimintaa, tai vain metsätaloutta, vaan jotta tämä haja-asutusalueiden yritystoimintaa halutaan ylläpitää, niin se täytyy ajatella sillä tavalla että ne on monitavoitteisia yrittäjiä että ne hankkivat sen tulonsa monista eri tulonlähteistä vuodenkierron mukaan' (Forest Owners Union in Northern Finland).

56 Original text: 'mahdollisuuksia tarjottavana jos metsätalouden tuotto vähenee' (Common Forest II, Finland; cf. Common Forest I, Finland).

57 Original text: 'eihän venäjän puu ole sertifioitua että kun ne kertoo että asiakkaat siis ihan päivittäin sinne tulee kyselyjä että mikä on puun alkuperä ja minkälaisesta metsistä ne on hakattu' (Lapland Forest Centre, Finland).

58 Original text: 'vi funderar ju mycket hur vi ska gå fram med att göra den delen ... det är ett ganska stort marknadskrav från framför allt England' (Sawmill Owner, Sweden).

59 Original text: 'alla virkesköpande bolag i stort sett är ju certifierade så att de har en högre nivå än skogsvårdslagen' (Common Forest, Sweden).

60 Original text: 'Det blir ju moment tjugotvå: man får tillstånd att avverka men man får ingen som ska köpa virke ... I och med att skogsvårdsstyrelsen lämnar tillstånd till avverkning så ... får man inte ... ersättning från skogsvårdsstyrelsen' (Common Forest, Sweden; cf. Norrbottens Läns Skogsägare, Sweden). This is also pointed out by private forestry interests in Sweden (which is PEFC-certified, but where the PEFC standard has moved rather close to FSC): 'many forest owners may fall into such a dilemma, a trap in some way ... most are starting to be certified now, all companies, organizations and buyer sawmills ... and they cannot accept this [timber]' [Original text: 'många skogsägare kan ju hamna i sånt dilemma, sån där rävsax på något sätt ...

de flesta börjar ju vara certifierade nu, alla bolagen, föreningarna och köpsågverk ... då får man inte ta emot det här [timret]' (Norrbottens Läns Skogsägare, Sweden)]. An interviewee involved in private forestry also argued that voluntary environmental protection should be emphasized, and even that voluntary environmental protection by private actors should be valued as highly as environmental protection measures imposed by the state (Norrbottens Läns Skogsägare, Sweden).

61 Original text: 'blåser de upp det då i media att det avverkas urskogar ... då törs inte skogsbolagen ... köpa den skogen' (Common Forest, Sweden).

62 Original text: 'uppbyggt efter ett skogsbruk där ju bara har brukad skog ... och där det inte är avsatt så mycket reservat sedan tidigare. Det är ju inte alls avpassat efter inlandet här' (Common Forest, Sweden; cf. Sveaskog, Sweden).

63 Original text: 'har du en omloppstid på 130 år så är ju inte 160 år så gammal skog' (Common Forest, Sweden).

64 Original text: 'certifieringen det är ju en bra grej men den är ju inte utformad så att den passar just då våra skogsmarker' (Common Forest, Sweden; cf. Forestry and Reindeer Herding Administration, Sweden).

65 Original text: 'det är bara miljön som väger ... de andra två benen [socialt och ekonomiskt hållbart skogsbruk] de är bara stödben' (Common Forest, Sweden).

66 Original text: 'sociala värden de är ju att man kan bo kvar på sin fastighet. Att man kan ha utkomst från skogen därför att det finns skogsentreprenörer och åkare och tillsammans så blir man så många att man kan ha kvar affären i byn och skolan och så vidare ... Det är ju sociala värden det också. Gör man avsättningar utav den mesta skogsmarken då stannar ju det ekonomiska flödet i byn och då blir det ju avfolkat' (Common Forest, Sweden).

67 Original text: 'det är kundkraven som har slagit igenom ute i Europa' (SCA, Sweden).

68 Original text: 'Jag tror att på sikt så kommer det att vara någon form av certifiering på det mesta skogsbruk. Det är ju ett krav från konsumenterna för att de ska kunna säga till sina kunder att det här kommer från ett skogsbruk som sköts på ett visst sätt ... alltså de här stora byggfirmorna' (Common Forest, Sweden).

69 Original text: 'vi är ju ... certifierade, vi har inte märkt av att marknaden skulle ha några andra planer ... men det är ju inte tillräckligt, det ska ju ändå avsättas mer skog' (SCA, Sweden).

70 Original text: 'jatkuvasti ollaan yhteistyössä ... poromiesten, kunnan, maanviljelyn eli kaikkien kanssa on yhteistyö, metsästäjien, Metsähallituksen' (Common Forest II, Finland; cf. Common Forest I, Finland).

71 Original text: 'oftast blir det ju man pratar om vad har kommunerna för framtid man pratar om att man ska utbilda ... man pratar inte i de här industritermerna vad är viktigt att det finns någon utveckling kvar i de små och medelstora företagen i träbranschen det nämns ju inte' (Sawmill Owner, Sweden).

72 Original text: 'det finns en verklighet som de inte har en aning om det problemet kan finnas' (County Forestry Board, Sweden).

73 Original text: 'skogsnäringen har gles befolkning' (County Forestry Board, Sweden).

74 Original text: 'Det har skett så kolossalt mycket inom organisationerna ... [idag är vi] egentligen enskilda individer som jobbar med [olika frågor]' (Forestry and Reindeer Herding Administration, Sweden).

75 Original text: 'vi vet vem vi ska prata med' (County Forestry Board, Sweden).

76 Original text: 'blir lite för långt… för långa avstånd mellan de här ställena' (Forestry Machine Driver, Sweden).

77 Original text: 'skogen är ju som en allmänning … om du som privatperson vann två miljoner på lotto eller någonting och så placerar du det i aktiekorgen så fick du förvalta den precis som du vill … om du köper ett skogshemman för två miljoner så har allmänheter och miljögrupper och allting har ju åsikter om hur du förvaltar ditt capital' (Common Forest, Sweden).

78 Original text: 'Ihmisiä on paljon Lapin läänissä ja ympäri Suomea niin ne ei ymmärrä, niillä ei ole harmaata aavistustakaan mikä on yhteismetsä … se on monelle hämärä käsitys' (Common Forest I, Finland).

79 Original text: 'kylmä että ennen kuin lunta sataa että kunnolla jäätyisi maa ja tuommoiset talvitien pohjat, että sehän se kaikista paras on … Ei ole liukasta ja tiet kestää ja esimerkiksi jos täytetään joku oja niin se jäätyy ja niin päin pois, että se se on kaikista paras' (Transport Entrepreneur, Finland; cf. Common Forest, Sweden).

80 Original text: 'Om det är en kall höst och ser ut att bli vinter och kallt och fruset och allt det där då går vi in på svagare marker för tidigt och så slår det om det blir liksom tjälförändring släpper i marken då måste vi ju byta om till starkare marker' (Sveaskog, Sweden).

81 Original text: 'talvellakin pikkupakkanen joku 10–20 on aivan ihanteellinen' (Transport Entrepreneur, Finland).

82 Original text: 'kuljetuspuolen kannalta, niin se on varmaan huonoa. Ensinnäkin liukkaat kelithän lisääntyy ilman muuta siis periaatteessa maantielläkin' (Transport Entrepreneur, Finland).

83 Original text: 'sista fem sex åren har vi nästan inte haft någon snö före jul och sedan är det en stor blida efter jul och då har det nästan varit barmark igen' (Sveaskog, Sweden).

84 Original text: 'vintrarna har varit mildare än vad de har brukat vara vanligtvis … ibland har det nästan varit nästan så att vi inte haft vit jul här heller och det har man ju alltid brukat ha' (Norrbottens Läns Skogsägare, Sweden).

85 Original text: 'tämä lauhtuminen näyttää lyhentävän talvea, mutta lumisuuttahan se ei meiltä poista' (Lapland Regional Council, Finland).

86 Original text: 'det har blivit grått tråkigare på många sätt [att arbeta]' (Forestry Machine Driver, Sweden; cf. Sveaskog, Sweden).

87 Original text: 'Fast det är ju optimalt jättebra i skogen nu jättehårt och fint frysen mark och lite snö [och] lätt att köra [på] vintervägar' (Forestry Machine Driver, Sweden).

88 Original text: 'i regel har ju ja den här tiden på året har det har blivit varmare så har det ju blivit mer regn än förut … och vinden börjar ju blåsa från västerut … snön den landar i fjällen' (Forestry Machine Driver, Sweden).

89 Original text: 'Idag alltså så har vi ju inte så kraftiga blidor att man måste ställa om avverkningar. Men å andra sidan så har vi ju kostnader på vägsidan då med väghyvlar och att sanda och sådär på is efter vägarna så att lastbilarna kan komma fram. Det är ju viss merkostnad med sandning och hyvling och sånt där, men det är ju inte så kraftigt att man måste ställa om avverkningsplaneringen' (Common Forest, Sweden).

90 Original text: 'talvella pystytään käyttämään talviteitä, jotka ovat kustannuksiltaan paljon paljon halvempia' (Metsähallitus, management).

91 Original text: 'aika tapauskohtaisesti että lasketaan minkälainen on se puumäärä ja kuinka kaukana se on siitä normaalista kesätiestä. Se täytyy ihan kustannuslaskelmalla

tehdä, että kannattaako meillä nyt odottaa että ottaisimme talvella' (Metsähallitus, management). It is noted, however, that even if the climate changes drastically from today's, more snow – or at least having a snowy season – may support other sectors, such as tourism: 'Warmer weather seems to be shortening the winter but it won't take the snow away. But it might take it away elsewhere and even improve the demand for winter tourism here' [Original text: 'lauhtuminen näyttää lyhentävän talvea, mutta lumisuuttahan se ei meiltä poista, mutta se saattaa poista muualta ja sen kautta saattaa tämmöinen talvimatkailun kysyntä jopa parantua meillä' (Lapland Regional Council, Finland)]. The changes in the relative length of the seasons may negatively impact the situation compared with today: 'For car testing there could be problems. We have had some soggy autumns where the ground has not frozen until December ... then the season shrinks so the hotels that thought they had guests coming lose lots of bookings for a month perhaps and that is a lot of money' [Original text: 'För biltesteri kan det ju bli problem. Vi har ju haft några surhöstar där det inte har frusit förrän mot december ... då krymper alltså säsongen så hotellen som trodde de hade gäster i de tappar ju massvis med rum en månad kanske och det är ju det blir mycket pengar' (Local Government II, Sweden)]. In addition, when this happens, tourism entrepreneurs take on employees for shorter periods.

92 Original text: 'det är ju på vår vår och senhöst som man har problem med bärighet och i marken och på vägar och då är det ett begränsat utbud av virke som finns' (Common Forest, Sweden).

93 Original text: 'är du ute på vinterväg så påsk då ska du vara ute därifrån ... [det är en] gyllene regel' (Sveaskog, Sweden; Common Forest, Sweden).

94 Original text: 'i fjol så började det ju tina långt före påsk ... så det var ju panik på vissa ställen' (Common Forest, Sweden).

95 Original text: 'i och med att vi skulle få virket till dem för det finns ju som inga lager heller, sågverken har inga lager heller och skogsindustrin i stort har inga lager, så att det är en vecka från stubbe tills det sågas virke' (Common Forest, Sweden).

96 Original text: 'har vi problem med väderleken så har ju alla andra små leverantörer problem med väderleken och tillsammans så blir vi ju ett stort bortfall' (Sveaskog, Sweden).

97 Original text: 'köra på grus på vägen ...till stora kostnader ... det blir ju med våld man får köra upp innan det avverkningen går förlorad' (Common Forest, Sweden).

98 Original text: '[storbolag] har ju större vägnät ... så att de har ju lite bättre möjligheter att rusta sig i förväg ... en liten markägare är ju ofta beroende av grannens väg också' (Common Forest, Sweden).

99 Original text: 'om man då kostar på en dyrare väg till ett område för att kunna avverka det när det blir så tid så vill man ju inte göra avsteg på det när samebyn har ett annat intresse ... där det går så tar man hänsyn till varandras näringar men då har man ju en viss gräns att när det börjar kosta för mycket pengar då så ser man till sitt eget intresse istället' (Common Forest, Sweden).

100 Original text: 'ställen med snöglopp och sånt där det är ju ett bekymmer naturligtvis därför att det påverkar, hämmar, tillväxt' (Sveaskog, Sweden).

101 Original text: 'saadaan siemenvuosia useammasti ja metsän uudistuminen sehän helpottuu huomattavasti sitten' (Metsähallitus, planning).

102 Original text: 'Lämpösumma on justiinsa se mikä kasvua rajoittaa, eli se vaatii sen tietyn lämmön, vuoden keskilämmön' (Local Forestry Association, Finland).

103 Original text: 'Se jos 60 päivään tulee yksi päivää lisää niin se on enemmän kuin prosentti. Ja jos kuusi päivää tulee lisää niin metsän kasvu lisääntyy 10%' (Common Forest II, Finland).

104 Original text: 'vi [skulle] ju få snabbare virke men då skulle ju kvaliten försämras ... den norrländska furan ... har varit en stor marknadsfördel det har det varit norrländsk fura som säljer' (Sveaskog, Sweden).

105 Original text: 'Kanske man inte kan sälja tätvuxna furor på samma sätt då hamnar det ju mer på att man ska vara konkurrent med mellansydsvenska träd ... [och sälja till] Grekland ... där man uppskattar det här. Kanske man måste titta på priset' (Sawmill Owner, Sweden).

106 Original text: 'jos se tietyn järeyden saavuttaa, jos se saavuttaa vaikka 90 vuodessa' (Local Forestry Association, Finland).

107 Original text: 'Sitten jos vielä ... lämpiää niin se saattaa olla vielä parempi, nopeampi se kiertoaika vaikkapa 70 vuoteen' (Common Forest II, Finland).

108 Original text: 'No ainakin se hetkellisesti on. En tiedä mitä vaikutuksia sillä muuten on tuolla luonnossa, jos puu alkaa kasvaa nopeammin, että onko siinä muitakin vaikutuksia ... jotka syövät sitä puun kasvua' (Metsähallitus, planning).

109 Original text: 'se siihen vaikuttaa aivan ratkaisevasti, siihen teiden kuntoon. Ja tietenkin se on kustannus kysymys varmasti' (Metsähallitus, planning).

110 Original text: 'när vi har ... flera tusen kubikmeter efter någon väg och så slår det om på ett par dar ... ska det virket ligga då tills vägen har torkat upp ... då kan [vi] inte kan sälja det för att det är ... blånader och insektsskador och sånt' (Sveaskog, Sweden).

111 Original text: 'saattaa olla sitä laatu huonontumista tapahtua, eli sienitaudit lisääntyä' (Metsähallitus, planning).

112 Original text: 'då sade de att det berodde på att det var året före som var ett fuktigt år och så en ordentlig [varm] sommar året efter och då hade det brett ut sig verkligt stort ... sen så kan det ha berott på att det varit något, miljöförändringar' (Sveaskog, Sweden).

113 Original text: 'det är något vi har varit mer förskonade från här ... eftersom vi då inte får använda några kemikalier i Sverige idag va så är ju frågan vad vi skulle börja göra då [om vi skulle få mer insekter] ... Frågan är ju hur mycket varmare det måste bli för att vi skulle få' (Norrbottens Läns Skogsägare, Sweden).

114 Original text: 'om man planterar och om man sår så då gäller det ju att ha bra alltså bra plantmaterial och frömaterial så att man liksom skulle anpassa sig till det här. Och då skulle det vara viktigt att vi bedrev forskning så att vi då hade kunna erbjuda våra skogsägare rätt material som de sen ska plantera' (Norrbottens Läns Skogsägare, Sweden).

115 Original text: 'Pieniä sahoja meillä on mennyt konkurssiin, mutta niin meillä on aina, kuollut ja uusi tullut' (Lapland Regional Council, Finland).

116 Original text: 'då lägger vi ned inte är det väl nåt [annat att göra]' (Sawmill Owner, Sweden).

Chapter 4

1 These will thus not be described here but will be taken up for Norway in the chapter on fishing, where interviewees discussed the Norwegian Saami Parliament more frequently.

2 The Saami year is regularly defined in terms of eight seasons. Interviewees did not use the corresponding terms but instead described changes in terms of the four seasons. For this reason, and also for simplicity and comparability with the other chapters in the book, the descriptions will divide the year into the four conventional seasons.

3 Original text: 'då är det ju samlingar och skiljningar och slaktningar, under den där mörkaste sämsta tiden. Det är mest arbete ... i oktober, november, december' (Reindeer Herder I, Sweden).

4 Original text: 'minst femhundra renar ... [för att ha din] inkomst enbart av renskötsel' (Reindeer Herder I, Sweden).

5 Original text: 'kai se on nykyisin se kannattavuudesta. Porotalouden kannattavuus. Ei ole hyvä että se kustannusrakenne ja se tulorakenne ... että se huonolta näyttää. Kustannukset nousevat niin korkealle tulojen tuottojen suhteen, että siihen on monia syitä että porot on muuttuneet ja poronhoitajat ovat on muuttuneet, se koko systeemi missä poronhoito toimii on muuttunut ja se ympäristö' (Reindeer Research Centre, Finland).

6 Original text: 'den dåliga lönsamheten absolut det är problem nummer ett. Problem nummer två det är ju intrången här som sker från alla håll och kanter ... Det är ju inte bara skogsbruket som gör att betesmarkerna minskar utan det är ju alla andra intrång som läggs på varandra ... Det är flygplatsen, det är biltestverksamheten, det är turistverksamhet' (Reindeer Herder I, Sweden).

7 Original text: 'skräpskog som de sa själv skogsfolket för vår del var det ju inte skräpskog utan det var ju hänglavsskogarna här som försvann' (Reindeer Herder I, Sweden).

8 This focus differs somewhat from that seen in the literature on reindeer herding, and also to some extent in the author's newspaper survey, where arguments centred on the cultural continuity of reindeer herding as a Saami practice are emphasized alongside economic ones (cf., for example, *Lapin Kansa*, 19 December 2004).

9 Original text: 'Yrket ... har fått större krav på att få det att gå runt ekonomiskt ... mer jobb och högre omkostnader' (Reindeer Herder III, Sweden).

10 Original text: 'se on koneellistunut koko ajan ja teknistynyt koko ajan, kustannuksia muuttuu, ja sitten on tullut myöskin ruokinta ... se on lisännyt kustannuksia. Lääkintä poroille ja tämmöiset. Tietenkin siinä on positiivisia vaikutuksiakin, että ei ne pelkästään vie vaan ne tuo myöskin' (Reindeer Research Centre, Finland).

11 Original text: 'tvåhjulig motorcykel eller fyrhjuling ... det är nog ingen som går de där sträckorna numera som vi [gjorde] förut' (Reindeer Herder III, Sweden).

12 Original text: 'att de kunde bedriva rationellt ändå renskötseln det är ju på grund av att man har kunnat ta hjälp av en snöskoter ja rent av åka motorcyklar' (Reindeer Herder III, Sweden).

13 Original text: 'Kostnaderna de ökar ju hela tiden och framför allt på drivmedelssidan det har vi märkt, så är det ju allt annat som följer med – skotrar, flyg, helikopterkostnader' (Reindeer Herder I, Sweden).

14 Original text: 'äänen käyttöön poro tottuu enemmin tai myöhemmin' (Reindeer Herder II, Finland).

15 Original text: 'både sommar höst och vinter helst när det är lite snö så jag kunde inte köra skotern' (Reindeer Herder II, Sweden).
16 Original text: 'mitä sitten tehdään kun tämä tekniikka on kaikki kokeiltu ja jos siihen saumaan tulee vielä semmoinen ikäpolvi, joka ei osaa käsitellä poroja, eli ei osaa etsiä, ei osaa pitää koossa eikä osaa kuljettaa tai pitää koossa kuljettamisen aikana. Niin kyllä siinä varmasti joudutaan ihan totaalisiin vaikeuksiin' (Reindeer Herder II, Finland).
17 Original text: 'kan möte veggen hvis de ikke tar seg sammen og begynner å kutte ned på den teknikken og de der hjelpemiddlene' (Reindeer Herder I, Norway).
18 Original text: 'tarkoittaa poronomistajuuden keskittämistä' (Employment and Economic Development Centre, Finland)
19 Original text: 'Jag ser ganska mörkt på framtiden i rennäringen ... Ett stort problem det är ju återväxten utrymmet är ju begränsat ... som utvecklingen är nu går vi ju bara mot större och större företag, mer och mer renar, och så har vi ju en begränsande faktor det är ju betesutrymmet renen tar. Man kan inte ha hur många renar som helst. Och för att varje företag ... då ska hitta en nivå där man kan ha finna sin försörjning så kan man inte vara hur många företagare som helst och så kan man samtidigt inte vara för få heller för att kunna bedriva en effektiv renskötsel. Och så i det skedet få in unga friska människor som vill gå in och satsa i det här det är jättesvårt för det måste vara fruktansvärt tungt att börja ... Jag tror nästan man måste få ärva sig en renhjord' (Reindeer Herder I, Sweden; cf. Reindeer Herder III, Sweden).
20 Original text: 'ett fåtal företag går ju runt bara på rennäringen men många behöver kompletterande arbetstillfällen' (Reindeer Herder III, Sweden; Reindeer Herder I, Sweden).
21 Original text: 'da blir de jo på en måte en sentralisering av rettigheter' (Reindeer Herder I, Norway).
22 Original text: '[D]a betyr jo det at det er rettigheter som er tatt vekk fra de andre som har lyst til å drive med reindrift' (Reindeer Herder I, Norway).
23 Original text: 'poroelinkeinon osaltahan on tapahtunut tässä 20 vuoden aikana huomattava muutos sillä tavalla että aluehallintoviranomaisen työmäärä tässä on lisääntynyt. Sen voi kaikki kuvitella ja voit itsekin kuvitella että melkein kaikissa mitä elinkeinoon liittyy niin pitää olla mukana' (Employment and Economic Development Centre, Finland).
24 Original text: 'miljöfrågor fjällförvaltning jakt fiske skoterfrågor skogsvårdsfrågor ja det är ju förordning jaktförordnining och sånt' (Reindeer Herder III, Sweden).
25 Original text: '[D]et har ju skett en, jag säger en enorm förändring från det vi gick och åkte skidor på vintern så på så kort tid på tjugofem år' (Reindeer Herder III, Sweden).
26 Original text: 'eli siellä on ollut valtavia muutoksia viimeisten kymmenen vuoden aikana ja siellä osa poromiehistä yrittää vielä pitää siitä kiinni että poron pidetään luonnonlaitumilla tai ruokitaan maastoon, mutta osa on ihan selkeästi alkanut laittamaan porot tarhaan' (Reindeer Research Centre, Finland).
27 Original text: 'yli 30 kiloa olevat vasapainot eivät ole mitään erikoista' (Reindeer Herder I, Finland)
28 Original text: 'siinä on sitä muuta ruokintaa siinä' (Meat Buyer, Finland).
29 Original text: 'väldigt varierande' (Reindeer Herder III, Sweden).
30 Original text: 'det beror ju på vad för sammansättning du har på betesmarkerna ... de som har begränsade vinterbetesområden och har haft en väldigt hög avverkningstakt på

området [som vi] de kanske måste börja [med utfodring] tidigare' (Reindeer Herder III, Sweden).

31 Original text: 'särskilt på vårvintern och på våren som vi måste utfodra dem och även vid dåliga betesförhållanden på vintern också för nu finns det ju ingen gammal skog som de kan beta utav och ta den där laven' (Reindeer Herder III, Sweden).

32 Original text: 'som vi aldrig använde förr när vi kunde släppa renarna in i lavskogarna' (Reindeer Herder III, Sweden)

33 Original text: 'bristen på hänglavsbete på vårperioden har gjort att vi ofta får stödutfodra den här perioden då vi har flyttat upp. och då kan man säga att det där ger ju en direkt ekonomisk effekt för alltså det påverkar ju vårt egna kapital säg att vi får utfodra för 250 000 varje vår ... direkt ur den egna fickan' (Reindeer Policy, Sweden).

34 Original text: '[fodret] köper vi ... det är ju pellets det är ju det enda, allena saliggörande. Visseligen stödutfodrar vi även med ensillage ... det köper vi av jordbrukare ja i kustbygderna här' (Reindeer Herder I, Sweden)

35 Original text: 'även om man får det här katastrofskadeskyddet man får ju göra en egen insats där på femtio procent det är ju därför knappt man klarar av det' (Reindeer Herder I, Sweden).

36 Original text: 'Det er mange år siden vi fora. Vi fora da det var de dårlige årene ... og da fora vi [dem] også bare ute på fjellet. Vi hadde dem ikke i gjäre og sånt' (Reindeer Herder II, Norway).

37 A Norwegian reindeer herder also noted that Norway, like Sweden but unlike Finland, has a division into summer and winter grazing lands, which may result in better maintenance of lichen grounds (Reindeer Herder II, Norway).

38 Original text: 'Noen vil fore og andre vil ikke fore ... [så] vi [har] delt oss opp i forskjellige siidaer og det er like greit. Da får jo vi... bruke ganske store områder her og ... de som forer de bruker mindre områder' (Reindeer Herder II, Norway).

39 Original text: '[d]e fleste slår selv faktisk her hos oss. Det er folk som har litt eiendom på grunnen av fiskerettighetene ... jeg tror ikke vi vil komme til det at vi vil fore med pellets og kraftfor' (Reindeer Herder II, Norway).

40 Original text: 'Jeg tror det er forskjell hvis du forer dem i gjärde eller ute. I gjäde så blir dem litt som oppdrettsfisk. Samtidig så tör vi ikke å kjöre altfor mye på det. Vi tör ikke å gå ut aktivt å pröve sverte det kjöttet som er foret fordi vi vet at hvis de kommer mildvär i desember så må vi jo fore selv' (Reindeer Herder II, Norway).

41 In Finland, some private direct sale has been possible, but this is less of an option in Norway, largely because of the relatively high prices and benefits: 'You get so much support per kilo of reindeer meat from animals you slaughter that if you have fewer than 600 reindeer it is unprofitable to sell privately for NOK 65 per kilo' [Original text: 'du får så mye tilskudd per kilo rein du slakter så har du under 600 rein så var det ulönnsomt å selge privat til 65 kroner per kilo' (Reindeer Herder II, Norway)]. Private sale, with its more limited volume, is also less of an option in the large-scale reindeer herding culture in Norway, where direct sales of some 10 to 20 reindeer do not make a large difference in a herd of 2000 (Reindeer Slaughter House, Norway). There are, however, cultural reasons for keeping large herds that may be more difficult to change. Large reindeer herds traditionally confer status, and especially the older people involved in herding keep large reindeer herds throughout the year (Reindeer Herder II, Norway). This may be seen as limiting the adaptations possible through slaughter (The Reindeer Trade Authority, Norway). Also, technology has

made it easier for older reindeer herders to keep their herds instead of passing them on to younger generations when they no longer can work them (Reindeer Herder II, Norway). An option here might be to introduce changes in the system towards a quota stipulating the maximum number of reindeer or towards new regulations (Reindeer Herder II, Norway).

42 Original text: 'i konsumentled har det [renköttet] ju inte blivit billigare i butikerna' (Reindeer Herder I, Sweden).

43 Original text: 'ei ole tuota tehty yhtään kunnon tutkimusta että sitä olisi joku voinut verrata' (Reindeer Herders' Association, Finland).

44 Original text: '[v]eldig gunstige forhåll. Det er liten dödlighet, det er lite rovdyr, det er naturgitte ting som milde fine vintre … det har värt kroneår' (Reindeer Slaughter House, Norway).

45 Original text: 'prisstödet först och främst halverades vid årsskiftet. Dessutom har ju då priset … sjunkit katastrofalt' (Reindeer Herder I, Sweden).

46 Original text: 'det finns ju inga möjligheter att öka slaktuttaget för att kompensera det prisnedfallet. Man försöker ju ligga på maximalt varje företagare han slaktar ju maximalt och vad han bedömer är möjligt att ta ut av renhjorden. Överskattar jag då minskar man ju på kapital och det är ju inte hållbart i längden det heller utan då försöker man ju naturligtvis då med kompletterande intäkter' (Reindeer Herder I, Sweden).

47 Original text: 'monella paliskunnalla on semmoinen tilanne että heillä ei ole poroille mitään ostajaa tiedossa' (Reindeer Herder I, Finland).

48 Original text: 'kaikki me tehdään toistemme kanssa kauppaa ja … niin me sitten puhutaan niistä määristä suurin piirtein mitä on ja miten on kauppa käynyt' (Meat Processing, Finland; cf. Stakeholder meeting, reindeer herding, Sweden).

49 Original text: 'viime vuonna oli kartelliepäilykset' (Meat Processing, Finland).

50 Original text: 'Det er håplöst å konkurrere med det i pris' (Reindeer Herder II, Norway).

51 Original text: 'Se on meillä niin helppo nähdä kun menee tuonne pakastealtaalle jossa meillä on sekä saksanhirvikäristyspussit sekä porokäristyspussit niin arvatkaa minkä pussin se kuluttaja tavallisesti valitsee. Useat ottavat saksanhirveä ja ne ajattelee että se on vain käristystä' (Meat Processing, Finland).

52 Original text: '90-luvulla oli tuotanto alamaissa … ihmiset kulutuskeskuksissa niin ne unohtivat oikeastaan poroa ruokapöydästä ja juhlapöydästä ja tämä on varmasti ollut yhteinen ongelma poronomistajille sekä poronostajille että nyt on alettu tehdä jotakin sen eteen että ihmiset kuluttaisivat poroa että sehän se on se paras lääke … se on se varmin hinnantakuu' (Reindeer Herders' Association, Finland; cf. Reindeer Slaughter House, Norway).

53 Original text: 'Man klarer ikke å markedsföre idag … Vi må ha TV og vi må ha store aktörer som markedsförer og som kommer inn for å tjene. Det er ikke som för at folk kom hjem til deg for å kjöpe reinkjött. Det gjör dem ikke nå mer. Nå skal det väre i butikken og det skal väre sånne småforpakninger og det skal väre bruksanvisning på' (Reindeer Herder II, Norway).

54 Attitudes might pose a problem in such marketing: 'Americans do not eat reindeer because they are Santa's animals' [Original text: 'amerikanerne spiser ikke rein for det er Julenissen sine dyr' (Reindeer Slaughter House, Norway)].

55 However, in the present situation, some actors in reindeer meat processing may organize to buy meat and sell it cheaply for a year or so without establishing long-term

relations, thus lowering the market price and destabilizing the market (Reindeer Herder II, Norway).

56 Original text: 'jos se tömähtää näille markkinoille niin me ollaan kusessa suoraan sanottuna' (Meat Buyer, Finland).

57 Original text: 'Vi vet at prisene i Finland og Sverige har bare ramla og ramla hele tiden og så har vi hatt den tollbeskyttelsen, men nå hvis prisen faller i Finland eller Sverige enda mer så vil det lönne seg å importere med full toll og kanskje de gjör det allerede' (Reindeer Herder II, Norway).

58 Original text: 'da skyter jo den norske reindriftsnäringen seg selv rett i foten' (Reindeer Slaughter House, Norway).

59 Original text: 'Idag är det ju så att det är ju reglerat ... att du får ha ett visst antal renar punkt slut och det gäller att få ut maximal avkastning utav dem' (Reindeer Herder III, Sweden).

60 The ability to adapt is thus also dependent on the areas and characteristics of an area that each reindeer herder (or district or village) can use, which are often dependent on political rather than natural boundaries: 'All the districts lie at or above the highest determined reindeer number, and you do not have the possibilities for emergency grazing that you had before ... Earlier there was more room that you could take but now if you use emergency grazing land outside your own district, that happens at someone else's expense' [Original text: 'Alle distriktene ligger på det överste reintallet og over, og så har du ikke de mulighetene til krisebeite som du hadde för. ... Da var det mer plass å ta men nå hvis man bruker krisebeite som er utenfor sitt eget distrikt så går det på bekostning av noen andre' (Reindeer Herder II, Norway)]. This limitation on district sizes and limited availability of crisis grazing areas was also noted in Sweden (Reindeer Herder III, Sweden). If there are any problems, such as too much snow, when reindeer do not dig for lichen anymore but lose weight and require supplementary feeding, those who have areas with less snow or with arboreal lichen can move reindeer there, but 'those who have limited winter grazing areas and have had a very high level of logging in their areas may need to begin [supplementary feeding] earlier' [Original text: 'de som har begränsade vinterbetesområden och har haft en väldigt hög avverkningstakt på området de kanske måste börja [stödutfodra] tidigare' (Reindeer Herder III, Sweden)]. However, no particular adaptations to limit or change this situation of district boundaries were suggested, as it was recognized that current problems are more a result of limited resources for maintaining reindeer herds than of the boundaries themselves.

61 Original text: 'jämför man jordbruket ett normaljordbruk och normal renföretagare så har normaljordbrukaren fyra gånger högre stöd' (Reindeer Herder III, Sweden).

62 Original text: 'onko se nykyinen järjestelmä vaiko kiintiöjärjestelmä, niin kumpi niistä on parempi' (Employment and Economic Development Centre, Finland).

63 Original text: 'tämä ei voida tukien päälle rakentaa' (Reindeer Herder II, Finland).

64 Original text: 'Sametinget har jo da for så vidt selv värt skeptisk til omleggningen hvor man nå gör en relativt stor del av de ökonomiske virkemiddlene produksjonsavhengige, som innebärer selvfölgelig det at de som har de störste flokkene da får störste delen av tilskuddene' (Saami Parliament, Norway).

65 Original text: 'NRL har jo alltid värt mörkeblått og det er jo grunnen til at vi har fått sånne produksjonspremier, at man premier salg' (Reindeer Herder II, Norway).

66 Original text: 'hvis vi ikke får kontoll, vi får mange gode år og vi ikke får kontoll over reintallet pluss at markedssituasjonen ikke kan ta unna produksjonen så vil vi havne

opp i samme smörja igjen med et for höyt reintall og at man begynner å beite på kapitalen. Nå har vi kanskje utifra mine öyner kanskje beitet litt på kapitalen allerede' (The Reindeer Trade Authority, Norway).

67 Original text: 'Slik at så lenge man ikke har gjort noen ting med fordelningspolitikken og den type ting så vil det väre det eneste rasjonelle for den enkelte reineieren å öke flokken' (Saami Parliament, Norway).

68 Original text: 'i distriktsplanen skal ... stordistriktet definere at her er den siidaens vinterbeiteområde og den siidaen sin og den siidaen sin osv. Det skal väre klarlagt. Om den siidaen der ser at de har for mye rein så skal de ha en mulighet til å redusere å alikevel få nytte av den reduksjonen av at man får större areal per dyr. Det er det som er tanken med et stort distrikt ... å få vekk litt av den konkurransen' (The Reindeer Trade Authority, Norway).

69 Original text: 'rent samisk ... Stortinget ... tör liksom ikke å gjöre altfor grove grep ... eller tvangsgrep' (The Reindeer Trade Authority, Norway).

70 Original text: 'alle vil komme på 600 i tilfelle at kvota skulle bli satt [det höyeste reintallet man kan ha for å få tilskudd]' (Reindeer Herder II, Norway).

71 Original text: 'renantalet ... det begränsar liksom sig själv ... alltså det är ju ett begränsat utrymme på bete det går inte att ha hur mycket renar som helst och så enkelt är det. Alltså man kunde öka om det inte fanns något tak men det gör det och betesmarkerna krymper ju snarare kommer ju det där taket att minska' (Reindeer Herder I, Sweden).

72 Original text: 'betesmarkerna det styr ju renantalet ... det är ju helt och hållet betet som styr för det slår tillbaka, överbetar man så får man igen' (Reindeer Policy, Sweden; cf. Reindeer Herder III, Sweden).

73 Original text: 'för att varje företag företagare då ska hitta en nivå där man kan ha finna sin försörjning så kan man inte vara hur många företagare som helst och så kan man samtidigt inte vara för få heller för att kunna bedriva en effektiv renskötsel' (Reindeer Herder I, Sweden).

74 The issue of the exhaustion of pasture areas, and of how much this is a result of reindeer herding as opposed to other land uses, was not explicitly discussed by interviewees, although the question had been taken up in the press (see, for example, *Lapin Kansa*, 18 February 2000).

75 Original text: 'Hvis du ser reindrift i forhåll til andre näringer så har du en del press utenifra mot reindriften. Det er en av de störste trusselene mot reindriften per idag' (The Reindeer Trade Authority, Norway).

76 Original text: 'että näissä on näitä keskinäisiä jännitteitä eri elinkeinojen välillä kun on kyse samasta resurssista mitä hyödynnetään että kuka sitä käyttää ja miten käyttää ja mihin tarkoitukseen' (Reindeer Herders' Association, Finland).

77 Original text: 'Porotalouden osalta on suurin kysymys on se laidunten kysymys ja haluavat metsätaloudelle rajoituksia. Se on se perusasia' (Lapland Regional Council, Finland; cf. Reindeer Herder III, Sweden).

78 To some degree, larger conflicts over hunting also influence the forestry–reindeer herding relationship, although reindeer herders generally hunt themselves. The problem is that some hunting, for example by local residents, disturbs reindeer (Reindeer Herder III, Sweden).

79 Original text: 'regner vi med at halvparten av sommerbeitene faktisk er berört av de' (Reindeer Herder II, Norway).

80 Original text: '[v]i har fått leid inn advokat for vi klarer ikke dette selv. Vi ville ikke hatt en sjansje for at det er så store aktörer. Det er statskraft, det er miljökraft ... sånne aktörer ... [Vindmöller] er den största trusselen vi har' (Reindeer Herder II, Norway).

81 Original text: 'På den finske siden har du små campingplasser langs med hele elva, men på norske siden så har du ... [en] camping' (Reindeer Herder II, Norway).

82 Original text: 'turismen, vi ser at den kommer mer og mer. Det er jo en fremtidig utfordring' (Reindeer Herder II, Norway).

83 Original text: 'Så når norsk rein trekker over på reservatet [på russisk side] så skjönner ikke russerne ... hvorfor reinen kan beite der' (The Reindeer Trade Authority, Norway).

84 Original text: 'det största bekymret som jag ser det idag det är att inte länsstyrelsen som är naturvårdschef och rennäringen pratar skydd ... avsätter man flera hundra tusentals miljoner kronor till reservat så vore det ju önskvärt ... att man skulle kunna kombinera in på det här sättet' (County Forestry Board, Sweden).

85 Original text: 'När man bor då också så i ett sånt här litet samhälle vi vill ju inte på något sätt stoppa utvecklingen ... vi vill ju inte bo här alldeles ensamma, vi är ju lika beroende av infrastruktur och samhälle i övrigt som alla andra. Men jag är övertygad om att man kan göra saker och ting mycket bättre genom att i ett tidigt skede kanske inleda samtal till exempel med den sameby som är berörd hitta lämpliga områden och kanske samverka i högre grad än man har gjort' (Reindeer Policy, Sweden).

86 Original text: 'Dersom man får veldig få personer i reindrifta, med andre ord at man kjörer en rasjonalisering for å bli färre personer, så vil betydningen av reindrifta for den samiske kulturen bli mindre og argumentasjonen for å bruke reinbeitene og store areler til reindrifen vil bli svekket. Slik at det er ingen grunn til å, hva skal jeg si, å bånnlegge omtrent hele Finnmark dersom det er svärt få som lever av det' (Saami Parliament, Norway).

87 Original text: 'ILO ... lyfte den här frågan [om rättigheter] i ett helt annat perspektiv än man egentligen hade tänkt då vad gäller samerna som ursprungsbefolkning' (Forestry and Reindeer Herding Administration, Sweden).

88 This was especially the case in the Saami paper *Sågat* (for example the editions of 1 April 2004, 6 April 2004, 15 April 2004, 20 April 2004 and 19 June 2004). Mainly, arguments were put forward in support of the ILO convention to the effect that certain perceived infringements on Saami land use should be prohibited or hindered based on the convention and Saami customary land use in the areas concerned.

89 The sentiments discussed in this part are also to a large extent seen in newspapers, in, for example, editions of *Piteå-Tidningen* of 12 January 2001, 21 December 2001, 31 January 2002, 6 February 2002, 16 November 2002, and 15 April 2004.

90 Original text: 'det som också gör att det har blivit vind i ryggen på de här frågorna det är ju alltså förra veckan fick vi ju signaler från EU och även FN har ju varit tidigare också om att Sverige klarar ju inte av det här att hantera den här situationen bra' (Forestry and Reindeer Herding Administration, Sweden).

91 Original text: '[i] FNs permanenta forum för ursprungsbefolkningar ... är ordföranden ... [den samiske representanten] Ole Henrik Magga ... han hanterar alltså femtusen olika ursprungsbefolkningsgrupper runt om i hela världen ... och det gör ju att det blir våldsamt tryck alltså mot ja både finska och svenska och norska samer och till viss del också ryska' (Forestry and Reindeer Herding Administration, Sweden).

92 Original text: 'Det man ofta glömmer i Sverige va det är ju att renskötselområdet det är ju ungefär åtta miljoner hektar skogsmark ... De som lever och verkar inom det här om man tittar på skogens sida det är alltså 40 000 privata brukningsenheter plus storskogsbruket. ... Det ska du då jämföra med ungefär 2000–2500 samer. ... Där har du lite grann av den konflikthärd som finns ... det är väldigt många privata som är inblandade i det här' (Forestry and Reindeer Herding Administration, Sweden).

93 Original text: '[A]llt det här gör ju att det [rennäringsfrågan] är ju oerhört hett just nu, så på en kort tidshorisont så har det kommit väldigt tunga betänkanden och utredningar som har rört om i den här grytan' (Forestry and Reindeer Herding Administration, Sweden).

94 Original text: 'då blev det ju ett djävulens liv ... i norra Sverige om det här att nu blir det rättighetsförskjutningar emellan så kallade svenskar och samer och hur ska det gå och ska de få veto när det gäller brukandet av våra marker och så vidare då började diskussionen på allvar' (Forestry and Reindeer Herding Administration, Sweden)

95 Original text: '[rennäringspolitiska kommitteen] tycker i princip alltså i några förslag va att samerna ska ha inom citationstecken – jag tycker inte att man ska tolka det så men många tycker det – vetorätt när det gäller avverkning va. Och då kan man läsa deras förslag alltså rennäringspolitiska lite grann ja som man brukar säga att fan läser bibeln' (Forestry and Reindeer Herding Administration, Sweden).

96 Original text: '[H]otet är ju att det sitter folk nere på Rosenbad då som inte alls har med konflikten att göra, hur den fungerar alltså ... jag vet inte om syftet är att de ska få poäng i Bryssel eller vad det går ut på att de ska skriva på ILO- konventionen och inte har klart för sig om det har vad det får för konsekvenser' (Common Forest, Sweden).

97 Original text: '[de] samiska organisationerna vill ha mera rättigheter men det är även andra grupper i Västerbotten och Norrbotten anser att de ska ha mera ... jägare och organisationer och såna som säger sig ha skatteland' (County Administration, Sweden).

98 Original text: 'Norr- och Västerbotten ... finns en ortsbefolkning ofta ättlingar till samer långt bort i tiden som har slutat med renskötsel och som anser sig ha större rätt till de här markerna än staten har' (County Administration, Sweden).

99 Original text: 'de personliga relationerna med skogsnäringen är ju som bra med, det är ju de här övergripliga målen som vi inte är överens om' (Reindeer Herder III, Sweden).

100 Original text: 'De nya signalerna [i] FSC ... [är att] man tittar på hänglavsbärande skogar och ... [ska] väga in ... helheten för samebyarna när man pratar om hänsyn' (Forestry and Reindeer Herding Administration, Sweden).

101 See, for instance, *Lapin Kansa* of 15 October 2001 and 14 November 2004.

102 Original text: 'fellesarenaen for de som driver sammen er blitt färre en för og det har blitt större avstand imellom de som gjör at behovet for ordninger, eventuelt behovet for noe som så moderne som organisasjonsutvikling er det faktisk et behov for' (Saami Parliament, Norway; cf. Reindeer Research Centre, Finland).

103 Original text: 'Meillä pitäisi olla toimiva hallitus paliskunnassa ... [ja] hallitukseen pitää pystyä asettamaan vaatimuksia ... Koulutusta pitäisi hallituksen jäsenille järjestää, koska heillä ei ole sitä johtamiskulttuuria ... Niin sehän olisi reilu peli, mutta kun sitä ei ole täällä semmoista systeemiä' (Reindeer Herder II, Finland).

104 Original text: 'vi har samråd med turism, skogsbruk, länsstyrelser, kommuner, jaktvårdskretsar, fiskeområdeskretsar … vi har ju enormt mycket såna saker som vi måste hålla oss a jour med idag … samhällsplaneringen går så pass fort' (Reindeer Herder III, Sweden).

105 Original text: 'å forholde seg til fire kommuner og alt byråkrati er jo håplöst og samtidig så har ikke kommunene så mye beslutningsmakt heller. Ting går utenfor kommunen uansett hvis det er större saker, så det blir jo mer fylkesmannen vi har med å gjöre i saker som kjöring, barmarkskjöring, rovvilt og de större utbyggningsprosjektene, for kommunen er bare höringsinstanse og da går det videre til fylkeskommunen alikevel' (Reindeer Herder II, Norway).

106 Original text: 'Reindriftsforvaltningen er den ytre etaten for Landbruksdepartementet slik at det å skyte på Reindriftsforvaltningen er det samme som å skyte pianisten. Altså komponisten er departementet … trekker de stor linjene når det gjelder forvaltningen av reindriftsnäringen' (Saami Parliament, Norway).

107 Original text: 'fått massor av synpunkter men väldigt mycket synpunkter har berört de här rättighetsbitarna [skogsbruk – rennäring] som inte alls är länsstyrelsens nivå' (County Administration, Sweden)

108 Original text: 'poron hyvinvointi niin se on ennen kaikkea sääoloista johtuva tekijä' (Reindeer Herders' Association, Finland).

109 Original text: 'for reindriften er det jo best at sommer er sommer og vinter er vinter. Det er helt klart' (Reindeer Herder II, Norway).

110 Original text: 'helpot talvet, vähälumiset talvet, aikainen kevät ja ei sääskiä' (Reindeer Herders' Association, Finland).

111 Original text: 'det ble jo mildvär om hösten, tidlig på hösten og så fryste det og da var det dålig' (Reindeer Herder I, Norway).

112 Original text: 'Sånn som et har värt nå, sånne tempratursvingninger som går så fort – vi har minus 20 grader på morningen og så på ettermiddagen har vi 2 pluss – det sier alle at det ikke har värt så ustabilt för. … Det går jo fram til jul og i de siste årene så har man ikke hatt snö för til jul. … Det synes folk er veldig merkelig' (Reindeer Herder II, Norway).

113 Original text: 'jag minns då aldrig till att när jag började [för omkring trettio år sedan] att vi hade såna här töväder mitt i vintern … jag upplever att vi hade ett lite stabilare klimat' (Reindeer Herder III, Sweden; cf. Reindeer Herders' Association, Finland).

114 Original text: 'flyttar upp till fjälls alldeles för tidigt och får mycket sämre produktion och då sjunker lönsamheten fast du har stora kostnader' (Reindeer Herder III, Sweden).

115 Original text: 'Man må jo si igjen at tempraturene har värt mildere, men … det er en for kort tidsperiode å snakke om' (Reindeer Herder I, Norway).

116 Original text: 'On puhuttu ilmaston lämpenemisestä, että ilmasto alkaa pahimmassa tapauksessa lämmetä, mutta onko siitä kaksi tai kolme vuotta kun oli toista viikkoa melkein 50, semmoista 45–48 astetta pakkasta' (Local Government, Finland).

117 Original text: 'minkälainen talvi muodostuu riippuu pohjasta, eli kuinka äkkiä maa routaantuu ja kuinka paljon lunta tulee ja tietenkin se tuleeko siihen semmoista jääkerrosta' (Reindeer Research Centre, Finland).

118 Original text: 'det fryser före det snöar då brukar det alltid bli bra bete bra betesförhållanden. Och då får det ju inte komma nåt något milt väder så den tiden … Då

tinar inte backen upp så då blir det absolut kanonbete och då, kommer kall snö på backen sen, blir [det] bra' (Reindeer Herder II, Sweden).

119 Original text: 'uppehållsväder tills i oktober, helst i slutet av oktober vill jag att det helst ska börja snöa kall snö' (Reindeer Herder III, Sweden).

120 Original text: 'de [renarna] hinner ju äta upp efter hösten efter brunsten' (Reindeer Herder II, Sweden).

121 Original text: 'att samla för slakt och skiljning det har ju varit så kostsamt när det är lite snö' (Reindeer Herder II, Sweden; cf. Reindeer Herder III, Sweden).

122 Original text: 'idealiskt klimat [är] ungefär max femtio centimeter snö på vintern, stabilt vinterklimat, alltså tio upp till tjugo grader kallt' (Reindeer Herder III, Sweden; Reindeer Herder II, Sweden).

123 Original text: 'det får ju inte komma för mycket snö ... om det blir över en meter snö, en och tio ungefär, då slutar de gräva' (Reindeer Herder II, Sweden)

124 Original text: 'kylan den gör ingenting ... har de [renarna] bara bra bete och är i gott hull' (Reindeer Herder I, Sweden).

125 Original text: 'det kan bli låste beiter og is over hele vidda' (Reindeer Slaughter House, Norway).

126 Original text: 'Det verste er jo hvis det blir mildt og skifter så mellom ... regn og kulde' (Reindeer Herder I, Norway; cf. Reindeer Herder I, Sweden).

127 Original text: 'Reinene dör som fluer og da er det plutselig for lite rein. Det kan skje og det er naturen som styrer det der' (Reindeer Slaughter House, Norway).

128 Original text: 'det blir mer och mer sura höstar ... det är ömsomt tö ömsomt kallt ...och det blir dåligt bete ... det är mera regel än undantag' (Reindeer Herder I, Sweden).

129 Original text: 'Sista åren så har det ju varit förändringar, helt påtagliga förändringar... kraftiga töväder på vintern ... när klimatet ska vara stabilt – det ska inte vara i januari februari upp till sju, åtta, tio grader i solen' (Reindeer Herder III, Sweden).

130 Original text: 'i fjol till exempel så fick vi var vi nästan utan snö i mars månad sådant som aldrig ens har hänt' (Reindeer Herder III, Sweden).

131 Original text: 'I vinter så höll det på att förstöra betet totalt med det här tövädret som var bara för en vecka sen, så vi är ju mycket oroliga när det gäller väderleksförändringar' (Reindeer Herder III, Sweden).

132 Original text: 'Det finns väl en gräns där alltså om vintern blir så kort att man klarar det här om vintern skulle bli så kort, att den kanske bara blev tre månader, men det lär vi nog knappast få' (Reindeer Herder I, Sweden).

133 Original text: 'hvis vi får mer ising av beitene enn det vi har hatt för, hvis det er tilfelle at det blir varmere at det ikke bare er noe som er forbigående, så kan det endre driftsformen' (Reindeer Herder II, Norway).

134 Original text: 'enten at det vil bli mer basert på foring ellers så vil vi måtte kanskje minske reintallet' (Reindeer Herder II, Norway).

135 Original text: 'då blir det ohållbart att ha en livskraftig rennäring' (Reindeer Herder III, Sweden).

136 Original text: 'når det blir mildvär på vinteren og de får bedre beiteforhold [inland] ... så iser det hos oss' (Reindeer Herder II, Norway).

137 Original text: 'det som har värt bra de to siste årene er at det [nedisningen] har först skjedd i de to förste ukene i februar og da er det ikke så lenge til at det begynner og varme seg opp igjen på bakken og sola kommer og det blir barflekker i snöen' (Reindeer Herder II, Norway; cf. The Reindeer Trade Authority, Norway).

138 Original text: 'Det är ju positivt om den [snön] smälter tidigare. Ju tidigare vår det [blir] ju längre blir ju grönbetetssäsongen och ju kraftigare och finare renar får man' (Reindeer Herder I, Sweden).

139 Original text: 'att det inte börjar att förtöa ... [utan] att det börjar töa då i april månad ... och så sen att det då fryser mot nätterna och på dagarna så töar betet upp och sen när man går in mot maj månad då får det töa dygnet runt ... och bli bart så fort som möjligt' (Reindeer Herder III, Sweden).

140 Original text: 'kan [vi] driva renarna om nätterna och så beta om dagarna. Då är det lätt för renen att gå och så får han mat i sig på dagen ... det är jobbigt för de här dräktiga och kalvarna ... som får gå i tung ... sursnö' (Reindeer Herder III, Sweden).

141 Original text: 'ibland genom de här väderlekstyperna vi har haft ... [har] isarna ... varit [så] svaga att hjordar har gått ned sig ibland och alltså [man har] kört ned med skoter' (Reindeer Herder III, Sweden).

142 Original text: 'tvingas du köra upp [med lastbil]' (Reindeer Herder III, Sweden).

143 Original text: 'det kostar [att köra upp] men de som är lite folk ... kan det ju då vara hugget som stucket att köra upp' (Reindeer Herder III, Sweden).

144 Original text: 'sedan jag började har det ju aldrig hänt [att vi fått blötsnö som dödat kalvar] ... det verkar ju som att den skulle komma sommaren direkt och nu är det ju så extrema somrar varmt' (Reindeer Herder II, Sweden).

145 Original text: 'tuli vielä lunta kauheasti ja vasat kuoli' (Reindeer Herders' Association, Finland)

146 Original text: 'nämä sääolot nimenomaan niin alle 10 vuodessa ollaan oltu kahdessa ääripäässä. Tämä on erittäin mielenkiintoinen ilmiö pohtia' (Reindeer Herders' Association, Finland).

147 Original text: 'Idealiskt är att det är lagomt varmt under ett antal veckor på sommaren och nattkallt och så regnar det en gång i veckan eller något sånt där så att det kommer upp nytt gräs eller och sen att augusti är sval' (Reindeer Herder III, Sweden).

148 Original text: 'Sommaren ska ju vara lagomt varm den får ju inte vara för varm heller för då kan vi inte märka kalvarna för då är det ju då tar värmen en sån påfrestning på kalvarna' (Reindeer Herder III, Sweden).

149 Original text: 'då betar de ingenting då är det ju mygg ... och olika bremsar och irriterar dem hela tiden, och driver upp dem från de här bördiga områdena där det är mycket gräs och ris' (Reindeer Herder III, Sweden).

150 Original text: 'räddningen var då att det var så torrt så att myggen inte kläcktes' (Reindeer Herder III, Sweden)

151 Original text: 'det är ju inte bra för renen för att då snöfläckarna töar ju bort och gräset alltså torkar ju bort' (Reindeer Herder III, Sweden).

152 Original text: 'de får ju inte den tillväxt som de är i behov utav' (Reindeer Herder III, Sweden).

153 Original text: 'då hade vi ju tur i början av augusti ... kom det regn ... det blev ju jättemycket svamp [som renarna kunde äta]' (Reindeer Herder II, Sweden).

154 Original text: 'vi har ju sett det nu de här somrarna att det har varit väldigt lite vatten ... vattenflödet är inte normalt' (Reindeer Herder III, Sweden).

155 Original text: 'det får ju inte bli varmare [på sommaren] för de har ju huggit bort alla de här granskogarna' (Reindeer Herder I, Sweden).

156 Original text: 'Vi bor jo så nært havet ... så det er alltid fuktig her' (Reindeer Herder I, Norway).

157 Original text: 'våres rein får jo helt panikk ... Det kan väre enkelte värforhold en liten kort periode som det er mygg og de springer jo da som gale opp på de höyeste [bergs] toppene' (Reindeer Herder II, Norway).
158 Original text: 'sen on pakko sopeutua tai se kuolee pois' (Reindeer Herder II, Finland).
159 Original text: 'ihmiset porotaloudessa ovat vanhastaan semmoisia jo että hyviä aikoja seuraa huonoja aikoja... seuraa hyvät ajat' (Reindeer Herders' Association, Finland).
160 Original text: '[d]a må staten inn kanskje å betale for å få prisen ned ... og ... få opp forbruket' (Reindeer Slaughter House, Norway).
161 Original text: '[n]är slaktstödet från staten sänks får man spara in på annat, till exempel inte köpa ny skoter' (Reindeer Herder II, Sweden).
162 Original text: 'poromiehet ovat tehneet itse aika paljon ... mutta ... näissä vallitsevissa rakenteissa on se epätasapaino olemassa minkä vuoksi nyt on helppo esittää muulle yhteiskunnalle sekaantumista nyt sitten näihin asioihin positiivisesti' (Reindeer Herders' Association, Finland).
163 Original text: 'de reineierene som ikke följer reglerne får ikke statstilskudd' (Reindeer Slaughter House, Norway).
164 Original text: 'Reindrift vil nok väre så lenge som folk vil hålle på med det, men du kan få store omleggninger og du kan få store svingninger i lönnsomhet og hvor mange som kan livnäre seg av näringen' (Reindeer Herder II, Norway).
165 Original text: 'om då skogsbruket avverkar mycket, men är det ett gynnsamt klimat och sånt där, då kan det ju som kompensera det. Men går klimatet och skogsbruket och de här övriga störningarna också emot då försvårar det ju avsevärt' (Reindeer Herder III, Sweden).
166 Original text: 'I ett visst läge så ser man att näringen inte kan anpassa sig eller nu vet vi ju inte var den gränsen går, om vi är nära den' (Reindeer Herder III, Sweden).
167 Original text: 'utslagning' (Reindeer Policy, Sweden)

Chapter 5

1 Original text: 'Det var ikke alle som hadde båter til å drive med vinterfiske ... Nødvendigheten med å gjøre noe annet ved siden av ... da var det jo utmarka som gjaldt' (Coast Fisher II, Norway).
2 Original text: 'Du fisker samme kvantumet i dag med 15 000 fiskere som du gjorde med 60 000 før. En del ti år siden ... Det har vært ei rasende utvikling, ei teknisk utvikling' (Finnmarks Fiskarlag).
3 Original text: 'Disse primärnäringene har forsvunnet veldig fort i disse her nordligste kommunene som Tana og sikkert andre ... det har skjedd en sånn helt utrolig urbanisering' (Coast Fisher III, Norway).
4 Original text: 'I Finnmark så er det mere tradisjon rundt fjordfiske. Det har ikke värt kapitalintensivt og alle i området har hatt rettighet til å utnytte de maritime resurssene' (County Council, Norway).
5 Original text: 'Fiskeflåten var ikke i stand til å levere dem nok råstoff til å drive dem hele året, og da begynte man å bygge trålere' (Directorate of Fisheries, Norway).
6 Original text: 'man har hatt en kondemnering av båter ... man kjørte en strukturpolitikk for å få ned antall båter og bedrifter' (Directorate of Fisheries, Norway).

7 Original text: 'flyndre [and] kveita var ... relativt godt betalt, men den ble også veldig fort utfisket da når man begynnte med garn og veldig store båter som lå inn over fjordene' (Coast Fisher III, Norway).

8 Original text: 'Det største forandringene etter mitt syn er rettighetene. De håller på å forsvinne. Den størte ulykken ... det var kvoteordningen innenfor torskefiske. ... for fjordene' (Coast Fisher II, Norway).

9 Original text: '[Krabbe er] en pest og en plage og en velsignelse. For dem som har kvote så er det jo en meget bra inntekt på en måneds tid om høsten når det ellers hadde värt lite annet. ... Men for de som ror med garn om våren etter torsk eller rognkjeks så er det jo helt grusomt ... De får jo uhordelig mengde av ... ødelagt garn' (County Council, Norway).

10 Original text: 'Kongekrabben nå er så tallrik innen i fjorden at du ikke kan drive torskefiske med garn lenger og at du ikke kan drive fiske på rognkjeks med garn, så er det umulig å oppfylle det kvotekravet og da blir folk gående uten' (Fishing Interest Organization, Norway; cf. Coast Fisher III, Norway).

11 Original text: 'i tillegg til at resurssene var helt i bunn så var det en kjempe konkursbølge her i Finnmark' (Directorate of Fisheries, Norway).

12 Original text: 'gav fiskerinäringen nye bein å stå på ... sikret råstoff. Men i 1995–96 så var det en ny konkursbølge' (Directorate of Fisheries, Norway).

13 Additionally, actors to a lesser extent discussed the risk of overfishing and overuse of the limited resources, such as through 'unregistered re-loading' ['uregistrert omlastning'] on Russian ships (Finnmarks Fiskarlag) or whether overfishing has occurred in general, but this was in general taken up very little in the interviews (Norges Kystfiskarlag; Finnmarks Fiskarlag).

14 Original text: 'Det som har skjedd på 1990-tallet og på fiskerisiden er det at man har fått en helt annen omsettning av fisk. For det første, tidligere så var det forbudt å fryse fisk på båt men nå er det faktisk lov ... til å fryse ombord. Utover 1990-tallet så ble det ... etablert frysehotell ... fryselager for fisken. Man kan fryse fisken ombord og levere den til fryselager og levere videre til dem som byr mest penger, etter auksjons-prinsipp delvis' (Directorate of Fisheries, Norway).

15 Original text: 'Det var i de årene som det med marked kom veldig i fokus for da fant man ut at nå var det ikke kvantum som var problemet, nå var det prisene rett og slett. Og kanskje det at fileindustrien som var så tung, var ikke riktig så konkuransedyktig lengere' (Directorate of Fisheries, Norway).

16 Original text: '[D]en her satsingen på bedrifter som driver helårlig bedrift og at fiskeri skal på en måte väre en helårlig industri på lik linje med andre industrier, kanskje hadde litt for stor ambisjoner i forhold til at ... det [situasjonen] varier jo. Man har ikke greid å tilpasse seg varierende rammebetingelser godt nok' (Directorate of Fisheries, Norway).

17 Original text: 'Vi [hadde] en konkursbølge ifjor igjen ... noen bedrifter kom seg aldrig ordentlig etter 1995–96. De siste tre årene har man hatt en ganske kraftig nedtelling' (Directorate of Fisheries, Norway).

18 Original text: 'Det har blitt etablert turisme her om vinteren basert på skutertur og turer ut på fjorden for å fange din egen kongekrabbe ... Det er ingen annen plasser i verden som kan tilby dette' (Directorate of Fisheries, Norway).

19 Original text: 'det er stor forskjell [mellom gruppe 1 og 2] ... Jeg har en båt som er i gruppen mellom 8–9 meter og jeg har en torskekvote på 17–18 tonn ... og så har jeg

hyse og sei i tillegg. Den som er i samme båtlengde men i gruppe 2 [kvote] har vel enn 7–8 tonn torsk, så det er halvparten mindre' (Coast Fisher II, Norway).

20 Original text: 'hvis det er bedre pris og tilgjenlighet på hyse enn på sei så kan man ta litt mer ut av den og mindre ut av seien' (Directorate of Fisheries, Norway).

21 Original text: 'du klarer deg med den kvota til å forsvare det ... Men så har du redskap, at hvis det ikke følger redskap med båten så må du gjøre den investeringen og det er fort oppe i et par hundre tusen kroner. Men det er forutsettning at du får 50% tilskudd for å starte med fiske og da kan du forsvare det ... får du krabbekvote i tillegg så går du med godt overskudd. Men en som starter helt nytt han har ikke opparbeidet seg rettigheter til krabbekvote' (Coast Fisher II, Norway).

22 Original text: 'den eneste håpet for å komme seg opp i gruppe 1 kvote i torskefiske er å kjøpe en torskekvote, men det er jo ikke lønnsomt. Man må jo jobbe gratis i 10 år eller du jobber for banken med andre ord' (Coast Fisher I, Norway).

23 Original text: 'Før så arbeidet det til og med 4–5 [personer] per sjark, idag så er det knapt 2 personer. Normalt en fisker per båt. Rasjonaliseringen av fiske har gjort at färre personer kan få innkomsten sin fra fiske' (County Council, Norway).

24 Original text: 'det vi gjør, er å kapitalisere dette her. Vi bytter ut folk med penger hele tiden. Vi pøser inn penger, og kjøper oss teknologi, og må si opp folk på grunn av at vi har ikke råd' (Norges Kystfiskarlag).

25 Original text: 'Kompentansen forvinner og det blir disse oldringene igjen og når de stryker med så selger de kvotene sine som en pensjonsordning' (Saami Parliament, Fishing, Norway; cf. Coast Fisher I, Norway; Coast Fisher II, Norway).

26 Original text: 'har vært bortimot håpløst' (Norges Kystfiskarlag).

27 Original text: '[det er] overinvesteringer. Det blir investert i trålere og ringnotfartøyer, milliarder for året. Det er fisken som skal betale alt dette her. Det er investert rundt 8–9 milliarder de siste årene' (Norges Kystfiskarlag).

28 Original text: 'Bankene er veldig kortsiktig, de blir så jævlig redd hvis det begynner å gå galt med en gang, men vi vet jo den her næringen svinger. De har for lite tålmodighet' (Finnmarks Fiskarlag).

29 Original text: 'Det er blitt gode priser på tørrfisken etter hvert siden det er färre og färre som driver å henger fisk' (Coast Fisher I, Norway).

30 Original text: 'de har värt mest fleksible og har hatt midre faste kostnader, investeringkostnader' (Directorate of Fisheries, Norway).

31 Original text: 'Den store det er laksoppdrett. Det er jo viktig ... for de lokale samfunnene som har det' (County Council, Norway).

32 Original text: 'utover 1990-tallet ... produksjonen ble faktisk tidobblet på noen år og nå er vi oppe på enn 50 000 tonn [her i Finnmark]' (Directorate of Fisheries, Norway).

33 This was, however, not only seen as a result of farming but also due to problems of scale in production and variations in quality: 'In total, the quantity that is caught at the coast here in Finnmark and the fjords ... it is too small for you to be able to make a go of it and get something out of it ... additionally the quality of what is brought to the shore varies quite a bit' [Original text: 'totalt det kvantumet som tas opp på kysten her i Finnmark og i fjordene ... er det for lite at du kan organisere deg og få noe ut av det ... også så er nok kvaliteten på det som ilandføres av veldig varierende kvalitet' (Coast Fisher III, Norway)].

34 Original text: 'det er to faktorer som spiller inn på mi inntekt. Det er kvantum, og det er pris. Om jeg nå har, kan du si 35 tonn fisk i stedet for, skal en si 130 tonn på en enkel sesong ... så er det klart at vi må ha en bedre pris. Ut ifra det, så er filetindustrien lavpris hele tiden. Det er mengden som skal gjøre det hele, og ut ifra det med det kostnadsnivået vi har i Norge i dag, så er det rett og slett ikke lønnsomt ... med de lave kvotene som vi har' (Norges Kystfiskarlag).

35 Original text: 'Du kan garantere leveringstid, altså at det er kontinuerlig, kvalitet kan du garantere ... Du kan fore den [fisken] og da kan du ha den der i et år, altså veldig lang tid ihvertfall' (Saami Parliament, Fishing, Norway).

36 Original text: 'Internet har fått istand en närhet mellom fisker og marked, dem har blitt obs på hverandre. De har fått mulighet til å kommunisere på en helt annen måte en før' (Saami Parliament, Fishing, Norway).

37 Original text: 'Det [ligger] mye penger i den nye måten å tilpasse seg ... Det er private organisajoner som vil komme inn og trade ... og da en oppgående fisker som har kunnskap i det her og ... har en kvote, han vil ha mest mulig ut av kvoten ... han fanger den og har den i mära og han bruker sin fritid til det når han ellers ville väre ledig til å selge den' (Saami Parliament, Fishing, Norway).

38 Original text: 'hovedkluet her er rettighetene, for har dem ingen rettigheter så er det bare en saga blått, hele det der' (Saami Parliament, Fishing, Norway).

39 Original text: 'Hvis det fortsetter ... dagens politikk i 10 år til så er det ikke fiskeri her om 20 år. Ikke et menneske' (Coast Fisher III, Norway).

40 Original text: 'Folk i området her har drevet fiske her i området i uminnelige tider og Finnmark er jo det området i landet som har den eldste bosettningen. Det er det eldste registrerte etter C14 dateringer, 10 300 år er det på Magerøya. ... Hvis vi bruker utgangspunktet 10 300 år så har folk hatt rett til å utnytte resurssene i 10 285 år mens i det siste 15 år er det kommet eller man er blitt beskårret den retten' (Fishing Interest Organization, Norway).

41 Original text: 'man tidligere kalte det for 'åpen allmening' og det betyr at alle hadde rett til å fiske. Nå er jo den stengt, men det som skjer er jo det at man har fått et sånt godseiervelde innenfor fiskeriene som ser ut til bare å forsterke seg. At dem som er stor skal bli større. Det er närmest en politisk målsettning' (Coast Fisher III, Norway).

42 Instead, interviewees noted the need for local feedback to those who are in charge of managing the resources. One interviewee gave this example:

> *I remember when I was growing up as a boy ... this fjord was packed full of herring in the autumn ... for us children it was an impressive sight ... But it was not so for the adults who saw where this was going. I have later read the records from the fishing organization at the time, and there were so many protests sent from the local fisheries organization to Finnmark Fiskarlag, but no administrative authority wanted to stop the herring fishing that was going on ... When the herring was gone or fished out people were sitting again with an empty sea. The old people who had experienced all of this told us – this was before my time as a fisherman – that for many years there were no herring in the fjord during the winter so they had to go elsewhere to get cod and fish. It was the result of the resource administration of those days. [*Original text: *Jeg husker når jeg vokste opp som gutt ... så var denne fjorden her fullpakket met sill om høsten*

... for oss barn så var det et imponerende syn ... Men det var det ikke for de voksene folkene som skjønte hvilken vei dette her bar. Jeg har fått i ettertid lese datidens fiskarlagsprotokoller og det ble sendt så mange protester ifra det lokale fiskarlaget til Finnmarks fylkesfiskarlag, men ingen forvaltningsmyndighet ville stoppe det sillefiske som foregikk ... Da silden var borte eller oppfisket så satt folk igjen med tomt hav. De gamle folka som hadde opplevd dette her fortalte, dette var før min tid som fisker, at i masser av år så fantes det ikke torsk i fjorden om vinteren så de måtte andre steder for å få torsk og fisk. Det var resultat av datidens forvaltning.] (Coast Fisher II, Norway)

Similarly, actors noted that 'the places where you may fish with Danish seine in the fjords, those are very badly managed because the nets have become so efficient that if you have a fish population in a fjord, then today you take the entire population at one go' [Original text: 'de plassene som man får fiske med snurrevad i fjordene, det er veldig dårlig forvaltning for de har blitt så effektive nå at har du en gytebestand på en fjord så med de brukene idag så tar du hele bestanden iløpet av et kast kanskje' (Coast Fisher III, Norway)].

43 Original text: 'i den ytterste konsekvens kan et 20 talls fartøy ta hele den norske torskekvoten, kan en si. Det trenger jo ikke bo folk her ... men hvor trengs det å bo folk? Skal vi klumpe oss i sammen?' (Norges Kystfiskarlag; cf. Norwegian Raw Fish Organisation).

44 Original text: 'stort sett firmaer fra Vestlandet. De tar fisken ned ditt for de får litt bedre pris der nede ... industrien her i Finnmark de blir sittende der med skjegget i postkassa. Det er derfor alt går konkurs ... her i Finnmark' (Coast Fisher I, Norway).

45 Original text: 'det er jo sånn over alt, enten du ser på matvarekjeder, skal være større og større, de slår seg sammen, alt skal bare bli større og større. Det må bli stort nok, da blir det lønnsomt. Tidligere var det sånn at [du skulle bruke fiske før] ... et middel til å opprettholde [befolkning], sånn at det er ingen som snakker om det lenger. Vi har ord som nærhetsprinsippet ... Det blir ikke snakket lenger om det' (Norges Kystfiskarlag).

46 Original text: 'det er jo sånn overalt at hvis man har penger nok så får du det [kvoter] som du vil. Det har aldrig värt problem' (Coast Fisher III, Norway).

47 Original text: 'Det er klart at dess større enheter vi får dess mer dramatisk blir hver handel i fremtiden for områdene' (Coast Fisher III, Norway).

48 Original text: 'et politiskt stemningsskifte ... i en del av de partiene her i Finnmark kanskje først og fremst skyldes det at man ser at fiskeripolitikken også har värt feil i forhold til kyststedene ... man får en uttynning av bosettningen som føre til at i Finnmark om 20–30 år bor folk i Finnmark på bare noen ganske få plasser. Det er helt tydlig at fylkespolitikerne nå, at det iallefall er en utvikling som man ikke ønsker' (Fishing Interest Organization, Norway).

49 Original text: 'Vi har jo kjempet for de som bor der skal få lov til å fiske' (Saami Parliament, Fishing, Norway).

50 Original text: 'Alle som er innenfor området har like rettigheter ... når det gjelder å opparbeide rettighetene til området så kan du si at man kan tufte det på samisk historie, men når det gjelder [rettigheter] innenfor området så deler man likt' (Saami Parliament, Fishing, Norway).

51 Original text: 'prøver å fremme intressene til de mindre fartøyene som lander fisk lokalt og som skatter også til kommunene i området og holder områdene bofast.

Mindre båter betyr at det er flere som får sine inntekter fra fiskerinäringen' (County Council, Norway).

52 Original text: 'en måte å få rettighetene til å bli i området er å starte et resursselskap som kjøper opp kvoter og som leier ut kvotene til fiskere. Regionskvoter som blir forvaltet av fylke eller lignende organisasjon for å holde rettigheten til å fiske innenfor fylke' (County Council, Norway). However, some actors viewed such a policy at county level as a recent change, and there were critical voices saying that the county has not done enough: 'The Finnmark County Council has now changed the direction of its policies rather markedly, but until very recently the county did not have a policy that differed in any substantial way from what the Fiskarlaget or the central government wanted' [Original text: 'Finnmarks fylkeskommune har nå helt markant lagt om sin politikk, men helt frem til det aller siste så har fylkeskommunen ikke hatt noen vesentlig forskjellig politikk fra det som Fiskarlaget og det som de sentrale myndighetenen har ønsket' (Fishing Interest Organization, Norway)].

53 Original text: 'vi tenker oss at vi skal få en regional forvaltning der regionen har sine kvoter ... og utifra det kan drive rekrutering' (Saami Parliament, Fishing, Norway).

54 Original text: 'stor vekt på der er det her med rettsgrunnlag for å drive fiske for det er jo et punkt som er ganske bedrøvelig idag' (Fishing Interest Organization, Norway).

55 Original text: 'Tidligere var det større innblandning gjennom ekstraordinäre virkemidler og fylkeskommunen ... hadde mer virkemidler til den type ting ... nå har de ikke lenger det' (Directorate of Fisheries, Norway).

56 Original text: 'nå har jo Fiskeridepartetmentet kommet motvillig etter, men hvor langt det vet jeg ikke enda' (Saami Parliament, Fishing, Norway; cf. Coast Fisher I, Norway).

57 Original text: 'Det er ikke noe lokal eller regional politikk som dominerer den utviklingen her i disse näringene' (Local Government I, Norway).

58 Original text: 'fylkeskommunen har aldrig värt noe aktør i fiskeripolitikken' (Saami Parliament, Fishing, Norway).

59 Original text: 'Det er nasjonal politikk hele tiden som styrer folks handlinger ... [og] vi kjenner en stadig økt intensifiert utvikling hvor staten reduserer i inntektsoverføringer til kommunene ... det blir vanskeligere å ... skaffe inntekter her hos oss fremover' (Directorate of Fisheries, Norway).

60 Original text: 'nasjonalt ... er [det] sterke krefter som ønsker en kommunesammenslåing' (Local Government I, Norway).

61 Original text: 'statens eierskap til grunn i Finnmark [er] omstritt ... staten er forsiktig med å utnytte sitt eierskap til en større grad av privatisering av resurssene for at den rettighetsituasjonen er uavklart og staten vil ikke provosere' (Local Government I, Norway).

62 Original text: 'Det vi driver med er nasjonal forvaltning' (Directorate of Fisheries, Norway).

63 Original text: 'det er innført kraftige reguleringer på dem [de større fartøyene] i fjorden og det har jo värt innført begrensinger på snurrevad ... spesielt så vi håper at det kan hjelpe' (Directorate of Fisheries, Norway). With regard to regulation, interviewees were also dissatisfied with the regulations on the minimum boat size required to receive a king crab quota (Coast Fisher I, Norway). Some of the credit for the minimum length being reduced to six metres from the initial eight has been given to the Saami Parliament (Coast Fisher II, Norway). The issue of king crab requirements and regulation has

been more easily managed than many others, as the Fiskeridirektoratet's regional office has national responsibility regarding the king crab resource (Directorate of Fisheries, Norway). Administrators noted, however, that the minimum boat length stems mainly from safety concerns related to landing crabs on small boats: 'The reason why people have been sceptical of lowering it [the required length] is safety – pure and simple' [Original text: 'Grunnen til at man har värt skeptisk til å gå så ned det er rett og slett sikkerhet' (Directorate of Fisheries, Norway)]. Fishermen also noted that state support for fishing production is limited: 'Everything we get, we get from what we produce. Earlier we got various forms of support, too, for instance price subsidies and cheaper fuel for the boats, but now all of this is gone' [Original text: 'vi får alt gjennom produksjon. För var det jo sånn at det kom jo stötte til blandt annet pristilskudd og billigere besin som vi brukte til båtene, men nå er alt det der borte' (Coast Fisher II, Norway)].

64 Original text: 'Politikken i seg selv har hatt dårlig styring' (Coast Fisher II, Norway; cf. Fishing Interest Organization, Norway).

65 Original text: 'Stortinget som er uinteressert i lokalpolitikk, og skjer i ly av ren uvitenhet ... Det er lobbyvirksomhet det går på hele tida, og fiskebåtsredersforbundet er knallsterke der, og har penger og har fått en voldsom makt oppi dette her. De får noen politikere til å tro på dette her; at hvis det bare blir stort nok, skal det bli lønnsomt nok' (Norges Kystfiskarlag).

66 Original text: 'på grunn av at mange av de små båtene var misfornøyd med politikken til Norges Fiskarlag og mente at Norges Fiskarlag hadde for stor fokus på de store båtene ... dannet [de] et eget Kystfiskarlag' (Directorate of Fisheries, Norway).

67 Original text: 'Fiskarlaget har hatt en meget meget sterk innflytelse på norsk fiskeripolitikk gjennom veldig lang tid og nå ... [er] myndighetens politikk er så veldig mye innrettet på at det skal väre stordrift ... større og større enheter for de mener at det er det mest rasjonelle. Så nå faller både myndighetens politikk og Fiskarlagets politikk sammen i det her ønsket om større enheter' (Fishing Interest Organization, Norway; cf. Coast Fisher III, Norway).

68 Original text: '[fiskarlaget her i Tana] den er utdøende' (Coast Fisher II, Norway).

69 Original text: 'De har aldrig gjort noe for oss fjordfiskere' (Coast Fisher I, Norway).

70 Original text: 'gjør [det] at Råfisklaget som salgsorganisasjon kan ikke fungere sånn som den har gjort før ... [da] lokale fiskekjøperen var monopolist, men det er ikke der lenger' (Saami Parliament, Fishing, Norway).

71 Original text: 'vi har jo hatt stadige EU tillpassninger gjennom EOS, så man skulle tro at dette er ting som ville bli tatt opp der' (Directorate of Fisheries, Norway).

72 Original text: 'kanskje ikke hundre millioner men adskillige millioner kroner, 40–50 millioner kroner, det er klart at det ligger store bankinteresser i det hele' (Coast Fisher II, Norway)

73 Original text: 'Det er bankene som styrer egenlig, for det større lån du har ... så gir bankene signaler til departementet at 'vi må endre litt på det her for de skal overleve den gruppa der' ... 90 fotinger ... det var da den gruppen som bankene satset på. Det er bankene men fiskarlaget har også noe å si' (Coast Fisher II, Norway).

74 Original text: 'før stilte de ikke markedskrav og de samme kravene til lønnsomheten, men på lån og for så vidt på tilskudd også så stiller Innovasjon Norge nær sagt de samme kravene som bankene' (Finnmarks Fiskarlag).

75 Original text: 'Nord-Norge og spesielt i fjordene og delvis enkelt kyststrekninger også, det er jo samiske områder og det vil også si at de her urfolksrettighetene som

er nedfelt i foreksempel ILO-konvensjonen 169 [gjelder]. Selv om det har värt en diskusjon om gjelder ILO for sjø- og havområder, så er nok de fleste juridiske ekspertene enige om at den gjelder' (Fishing Interest Organization, Norway).

76 Original text: 'Man har [i Norge] ... tiltrådt det som finnes av konvensjoer for å sikre urfolks og minioriteters interesser, men det formelle lov- og regelverket det er sett helt bort ifra i denne sammenhengen' (Fishing Interest Organization, Norway).

77 See for instance *Ságat* of 20 January 2004, 29 May 2004 and 13 July 2004 and the references in Chapter 4 to the impact on reindeer herding of the ILO convention.

78 Original text: 'Det er mulig at det ikke er helt 100% rett men det er i värt fall aktørene i det så det vitenskaplige grunnlaget er selvfølgelig havforskning i de enkelte land' (Directorate of Fisheries, Norway). The concerns that can be seen in the research mainly pertain to the management of the limited quota: 'The way we have run it now it is entirely mad; we suddenly are allowed large catches and then we have a complete ban [on fishing for] two to three years ... But this is something that suits large fleets well, because they can take very much of a resource in two to four years ... but for all others ... who have to live from it, it is mad' [Original text: 'sånn som vi har drevet nå så er det helt vannvittig at vi har plutselig masse uttak og så har vi totalforbud i 2–3 år. Men det er noen som det passer veldig godt for og det er den store flåten for dem kan bare ta ut kjempe mye ut av en resurss der i en 2–4 år men for alle de andre som skal leve ut av det så er det helt hull i hode' (Saami Parliament, Fishing, Norway)].

79 Original text: 'det er jo Norges Fiskarlag og trålernäringa som er til stede da' (Saami Parliament, Fishing, Norway).

80 Original text: 'Alle de her fiskebrukene på kysten måtte ruste opp for enorme summer for å møte disse her EU kravene. Ellers så kjøper de ikke en eneste fisk' (Coast Fisher I, Norway).

81 Original text: 'Selvfølgelig i forhold til EU så er det med tollbarrierer på bearbeidet fisk ... en faktor som bidrar til at fisk fraktes ubehandlet ut av regionen Finnmark' (Fishing Interest Organization, Norway).

82 Original text: 'Forskere påstår at krabben ikke har noe innvirkning på flora og fauna i sjøen, men ... det som jeg og andre har registrert er at den her som vi kaller for 'bottenkontta' ... den er blitt borte' (Coast Fisher III, Norway).

83 Original text: 'det [er] ikke godt å si om det er noe fisk igjen her om ti år for tidligere så var det veldig mye av den her småtorsken ... men nå skal du väre uhyre godt kjent for å få det kokefisk faktisk på sommeren' (Coast Fisher III, Norway).

84 Original text: 'Hvis den [småtorsken] blir borte så er vi egentlig like langt og i tillegg så forsvinner faunaen så da forsvinner også andre arter med' (Coast Fisher II, Norway; cf. Norwegian Raw Fish Organisation).

85 Original text: 'Det er ikke bare krabben som er skyldig at det er lite fisk, for vi har også den her kråkebolleinvasjonen som har begynnt for en vel 20 år siden og som har beita tareskogen helt ned' (Coast Fisher III, Norway.).

86 Original text: 'Dermed har ikke småfisken mange plasser og gjemme seg ... den blir spist opp av skarv og fugl, sel og oter' (Coast Fisher I, Norway).

87 Original text: 'En del av fiskeslagene ser ut til å ha forandret vandringsmønster for at den her større seien kommer ikke inn på fjordene her som tidligere og torsk er det heller ikke mer på sommeren' (Coast Fisher III, Norway).

88 Original text: 'flyndra har forsvinnet og steinbitten har forsvinnet uten at jeg har noe god teori på hvorfor dem har blitt borte. Kobbenen renset godt i matfatet men det er

såpass lenge siden at dem skulle ha reetablert seg, men det kan jo ha noe med den her krabben iallefall for steinbitten vedkommende som gyter relativt grunt og samlet at de knasker den rognen i seg' (Coast Fisher III, Norway).

89 Original text: 'de lokale selene vi har, de trenger 5–6 kilo i døgnet fiskemat for å overleve og hvis det er noen tusen dyr i et område så skal det mye fisk til' (Coast Fisher II, Norway).

90 Original text: 'Det var dyrt selskinn, men nå er blitt sånn at det er en fy sak. Fiskeridepartementets hellig ku er selen' (Coast Fisher I, Norway).

91 Original text: '[Fiskeridepartementet] sier at det er ikke så mange som fiskerne påstår' (Coast Fisher I, Norway; cf. Local Government I, Norway).

92 Original text: 'det er en økologisk katastrofe men på grunnen av at vi ligger her og ikke 2 mil utenfor Oslo så er det ikke et problem. Er det et problem så blir det ikke tatt alvorlig' (Coast Fisher III, Norway).

93 Original text: 'store svingninger i bestandene som torsk, sei, hyse' (Saami Parliament, Fishing, Norway).

94 Original text: 'gjennom vårt reguleringssystem der vi fisker på visse årsklasser med visse metoder ... forsterker [vi] de svingningene som er i de enkelte bestandene' (Saami Parliament, Fishing, Norway).

95 Original text: 'man legger om til å fiske på ungfiskebestanden den her som er fra 4–7 år som er så her stor ... eller reguleringsåret, det tar jo utgangspunkt i første januar, hvis man hadde snudd det til 30 juni' (Saami Parliament, Fishing, Norway).

96 Original text: 'plutselig så begynner det å regne midtvinters og det er mild i 14 dager. Det er ikke normalt' (Coast Fisher I, Norway).

97 Original text: 'Vintrene her oppe i Finnmark ... begynner å minne meg mer og mer om de der Lofoten vintrene' (Norges Kystfiskarlag).

98 Original text: 'mildvær midt på ... vinteren, og lange perioder' (Norges Kystfiskarlag).

99 Original text: 'For 2 år siden så regnet og regnet det og hadde det forsatt å regne så hadde isen på Tanaelva gått' (Coast Fisher I, Norway).

100 Original text: 'i år så har det värt et helt ekstremt dårlig år rundt omkring i hele Finnmark. Det er ikke laks noen plass' (Coast Fisher I, Norway).

101 Original text: 'Gytinga må jo klaffe og det kan väre at det har muligens värt flom i elv når fisken gyter så om vinteren så fryser rogna seg fast i isen og blir ødelagt' (Coast Fisher I, Norway).

102 Original text: 'Det er som de gamlingene opp langs med Tana sier at det bør helst väre lite vann i elva når laksen gyter. Er det flom i elva så gyter de på de grusbankene som da blir tørre om vinteren' (Coast Fisher I, Norway).

103 Original text: 'Det kan väre at det er naturlige svingninger og det har det värt før at det har värt dårlige lakseår' (Coast Fisher I, Norway).

104 Original text: 'det har ikke blitt is på fjordene noen plasser i motsetning til før for 20–30 år siden når det var is langt utover i fjorden her ... hver vinter' (Coast Fisher I, Norway).

105 Original text: 'Tempraturen i sjøen har stiget med kanskje 1 grad eller 2, det sier forskerne ... tempraturen i Barentshavet er ganske høy mot det som det er normalt' (Coast Fisher I, Norway).

106 Original text: 'de siste vintrene ... har det ikke värt en eneste dag med skikkelig kuling. Vi har sluppet unna de värste vinterstormene nå flere år på rad' (Coast Fisher I, Norway).

107 Original text: 'mange fiskere, de stresser ikke slik i januar og februar, mørkeste og hardeste årstiden. For de vet at kvoten tar man opp allikevel' (The Norwegian Raw Fish Organization).

108 Original text: 'Her er det snakk om å få minst mulig utgifter for å ta opp fisken. ... Du ser på oljeforbruket, hva man bruker i olje for å få opp de ... tonnene' (The Norwegian Raw Fish Organization).

109 Original text: 'tidlig varme på våren som her på 1 mai eller iallefall før den 15. så var det opp i 20 grader. Det er jo helt ... unormalt og ifjor så var det noe lignende' (Coast Fisher III, Norway).

110 Original text: 'Det blir varmere temperaturer ... De siste par årene ... er [det] veldig stor fisk her, fisk som normalt har gått til Lofoten for å gyte stopper opp her utenfor Finnmark. Vi merker jo det at det er veldig mye rogn i den fisken ... levert i Finnmark fra fiskerne' (The Norwegian Raw Fish Organization).

111 Original text: 'torsken har begynt å gyte utfor Troms' (Norges Kystfiskarlag).

112 Original text: 'varmere vann og slik så vil du få en hurtigere produksjon. Rent umiddelbart ville det være positivt for mye. Torsken vokser bedre og gyter. Planktonet vokser og du får god næringshjelp' (Finnmarks Fiskarlag).

113 Original text: 'i og med at det ikke er slik det skal være så er det nok ikke positivt på lang sikt. Det fåes nok tilbakeslag på ett eller annet' (Finnmarks Fiskarlag).

114 Original text: '[Att det blir mer fisk] det kan jo godt hende det, men om ... det blir større kvote med det det vet vi jo ikke ... om det når ned til fiskern i båten' (The Norwegian Raw Fish Organization).

115 Original text: 'alger har [vi] faktisk hatt det et par tre somrer nå. Havet blir grønt, men det har ikke vært noe toneangivende problem, enda i hvert fall' (Norges Kystfiskarlag).

116 Original text: 'Det kan jo ødelegge mye for markedet og for å få folk til å spise fisk og slikt ... en [gang] på 1990 tallet ... [hadde] seien ... kveiste, det var jo stopp over døgnet på hele tyske markedet' (The Norwegian Raw Fish Organization).

117 Original text: 'Vi har jo ikke trengt flåten til å fiske makrell og sild ... Det skal jo kapital og alt slikt også tillatelse ... du må jo ha kvote i makrell. Der må du ha stor pengebok hvis du skal inn i silden og makrell fiske' (The Norwegian Raw Fish Organization).

118 Original text: 'makrellen kommer lenger og lenger nord. Den har vi jo faktisk allerede oppi Vestfjorden. Det var jo aldri snakk om makrell da jeg bodde der nede på 70 tallet, som jeg vet om i hvert fall' (Norges Kystfiskarlag).

119 Original text: 'hvis at fisken av temperaturmessige grunner ikke kommer opp til kysten så har ikke sjarkflåten noe sjanse' (Finnmarks Fiskarlag).

120 Original text: 'jo høyere temperatur ... så er det mer urolig vær' (The Norwegian Raw Fish Organization).

121 Original text: 'nå er det så små kvoter. 99% av fiskerne tar jo opp kvoten sin uansett' (The Norwegian Raw Fish Organization).

122 Original text: 'hvis man snakker om fiskerne hvordan de anpasser seg ... så er det på en måte to strategier i utgangspunktet. Den ene strategien er den som de fleste som er, litt yngre fiskere og skal bli i næringen. De kjøper seg opp, altså de investerer i kvoter for å få et bedre grunnlag. Så har du de som er... jeg vet ikke om det er noen strategi eller mangel på strategi: de gjør ingenting, de regner med at de lever med det her den tiden de har igjen' (Finnmarks Fiskarlag).

123 Original text: 'Finnmark flåten er ikke bygd opp til å drive ute i havet. Det er jo en kystnær flåte som er tilpasset kysten' (The Norwegian Raw Fish Organization; cf. Local Government I, Norway).

124 Original text: 'Du kommer til et punkt hvor det blir en fysisk hindring' (County Council, Norway).

125 Original text: 'Du har … ikke redskaper dimensjonert etter det, det er jo større dyp og strøm. Du … har jo ikke … det her fortøyningsutstyret for garn og ikke garn tellet som er dimensjonert for det. … Di kan liksom ikke omstille deg så veldig lett, det er en det ene etter det andre' (Finnmarks Fiskarlag).

126 Original text: 'du har jo en filetindustri her som er bygd på torsk, med maskiner som er bygd til torsk. De kan ikke skjære makrell, eller andre fiskeslag' (Finnmarks Fiskarlag).

127 Original text: 'vi har vel ikke de samme tradisjonene … de som har hengt tørrfisk og transportert den i Lofoten i generasjoner … de har inne en teknikk altså for å utøve selve den prosessen … og pleie importørene i Italia' (Finnmarks Fiskarlag).

128 Original text: 'det er nok et resultat av temperaturer. For å få prima kvalitet er du jo veldig avhengig av temperatur. … det kan fort surne hvis det blir mildt og slikt' (The Norwegian Raw Fish Organization; cf. Finnmarks Fiskarlag).

129 Original text: 'Om våren når vi har mye nedbør, store elver, det er veldig dårlig fiske da. Det er en periode om våren som mye ferskvann blir i sjøen, veldig stor oppblåsing av plankton. Det er veldig dårlig fiskeperiode' (The Norwegian Raw Fish Organization).

130 Original text: 'det er det dårligste [året jeg har hatt]. Jeg har aldrig fisket så dårlig med laks' (Salmon Fisher, Norway).

131 Original text: 'det var opp til 20 varmegrader i elva … Hvis vannet blir stadig varmere mot land så søker laksen seg mot de dype hullene' (Salmon Fisher, Norway).

132 Original text: 'Man kan ta en 3 år tilbake så begynte det å bli mindre, men det har vel mye å gjøre med det vannet at det har blitt så varmt' (Salmon Fisher, Norway).

133 Original text: 'det er det uten tvil at for når fisken seig seg ned dypere om vinteren i Januar, og i Februar spesielt og så var det jo sånn at mange hadde ikke båter til å legge seg lengre ute i Tanafjorden' (Coast Fisher II, Norway).

134 For instance, interviewees noted that 'the quota system is devastating this area' [Original text: 'kvotesystemet ødelegger for dette området' (Coast Fisher II, Norway)].

135 Original text: 'kystfiskerne har samme problemer over alt, altså rundt Atlanteren. Enten Portugal, Spania, Færøyene eller Island, de har akkurat de samme problemene, og vikende front på grunn av store kapitaler interesse for det og store handelsvarer' (Norges Kystfiskarlag).

References

I. Primary Material: Interviews

Forestry

Älvsbyhus, Sweden, Älvsbyn, 9 December 2003
Common Forest I, Finland, small municipality in Eastern Lapland, 15 March 2004
Common Forest II, Finland, small municipality in Eastern Lapland, 15 March 2004
Common Forest, Sweden, inland municipality, 27 February 2004
County Administration, Sweden, Jokkmokk, 17 December 2003
County Forestry Board, Sweden, Piteå River Valley District Administration, 10 December 2003
Forest Owners Union in Northern Finland, Rovaniemi, 31 March 2004
Forestry Machine Driver, Sweden, inland municipality, 11 December 2003
Lapland Forest Centre, Finland, Rovaniemi, 4 March 2004
Lapland Regional Council, Finland, Rovaniemi, 27 February 2004
Local Forestry Association, Finland, small municipality in Eastern Lapland, 8 March 2004
Local Government I, Sweden, Älvsbyn, 10 December 2003
Local Government II, Sweden, Arvidsjaur, 25 February 2004
Local Government, Finland, municipality in Eastern Lapland, 3 March 2004
Metsähallitus, management, Kemijärvi, 3 March 2004
Metsähallitus, planning, Savukoski, 25 February 2004
Norrbottens Läns Skogsägare, Sweden, 20 October 2004
Sawmill Owner, Sweden, inland municipality, 16 December 2003
Sawmill, Finland, Rovaniemi, 23 February 2005
SCA, Sweden, Piteå, 15 December 2003
Stakeholder meeting, forestry, Finland, Salla, 16 November 2004
Stakeholder meeting, forestry, Sweden, Arvidsjaur, 19 October 2004
Sveaskog, Sweden, inland municipality, 11 December 2003
Swedish Wood and Tree Trade Union, inland municipality, 11 December 2003
Transport Entrepreneur, Finland, municipality in Eastern Lapland, 8 April 2004

Reindeer herding

County Administration, Sweden, Jokkmokk, 17 December 2003
Employment and Economic Development Centre, Finland, Rovaniemi, 14 May 2004
Forestry and Reindeer Herding Administration, Sweden, within the National Board of Forestry, 1 March 2004
Meat Buyer, Finland, municipality in Eastern Lapland, 20 September 2004
Meat Processing, Finland, municipality in Eastern Lapland, 20 September 2004
Reindeer Herder I, Finland, municipality in Eastern Lapland, 20 September 2004
Reindeer Herder I, Norway, coastal area, 6 June 2004
Reindeer Herder I, Sweden, inland municipality, 26 February 2004
Reindeer Herder II, Finland, municipality in Eastern Lapland, 11 March 2004
Reindeer Herder II, Norway, coastal area, 21 December 2004
Reindeer Herder II, Sweden, inland municipality, 26 February 2004
Reindeer Herder III, Sweden, inland municipality, 28 February 2004
Reindeer Herders' Assocation, Finland, Rovaniemi, 24 October 2004
Reindeer Policy, Sweden, individual with policy interest in reindeer herding, inland municipality, 26 February 2004
Reindeer Research Centre, Finland, Kaamanen, 16 April 2004
Reindeer Slaughter House, Norway, Tana area, 17 June 2004
The Reindeer Trade Authority, Norway, Karasjok, 22 September 2004
Saami Parliament, Norway, Karasjok, 23 June 2004
Stakeholder meeting, reindeer herding, Finland, Rovaniemi, 23 November 2004
Stakeholder meeting, reindeer herding, Norway, Karasjok, 18 January 2005
Stakeholder meeting, reindeer herding, Sweden, Arvidsjaur, 18 October 2004

Fishing

Coast Fisher I, Norway, Tana area, 8 July 2004
Coast Fisher II, Norway, Tana area, 29 June 2004
Coast Fisher III, Norway, Tana area, 30 June 2004
County Council, Norway, Vadsö, 1 July 2004
Directorate of Fisheries, Norway, northern Norway area, 22 June 2004
Finnmarks Fiskarlag, Finnmark municipality, 17 January 2005
Fishing Interest Organization, Norway, Tana area, 7 July 2004
Local Government I, Norway, Tana area, 21 June 2004
Local government II, Norway, Tana area, 30 June 2004
Norges Kystfiskarlag, Finnmark municipality, 20 January 2005
The Norwegian Raw Fish Organisation, Finnmark municipality, 19 January 2005
Saami Parliament, Fishing, Norway, Karasjok, 5 June 2004
Salmon Fisher, Norway, Tana area, 23 September 2004
Stakeholder meeting, fishing, Vadsö, 11 March 2005
Stakeholder meeting, fishing, Vardö, 19 January 2005

References to newspapers

Lapin Kansa, 18 February 2000
Lapin Kansa, 15 October 2001
Lapin Kansa, 14 November 2004
Lapin Kansa, 19 December 2004
Piteå-Tidningen, 12 January 2001
Piteå-Tidningen, 21 December 2001
Piteå-Tidningen, 31 January 2002
Piteå-Tidningen, 6 February 2002
Piteå-Tidningen, 16 November 2002
Piteå-Tidningen, 15 April 2004
Piteå-Tidningen, 16 April 2004
Ságat, 20 January 2004
Ságat, 1 April 2004
Ságat, 6 April 2004
Ságat, 15 April 2004
Ságat, 20 April 2004
Ságat, 29 May 2004
Ságat, 19 June 2004
Ságat, 13 July 2004

II. Secondary Material: Published Sources

Aasjord, B. (2002) 'Where have all the fishes gone? Men shaping the marine arctic future', in Finnish Ministry of Social Affairs and Health (2002) *Taking Wing. Conference Report. Conference on Gender Equality and Women in the Arctic, 3–6 August 2002, Saariselkä, Inari, Finland*, Reports of the Ministry of Social Affairs and Health, No 12, Helsinki

ACIA (Arctic Climate Impact Assessment) (2004) *Impacts of a Warming Arctic: Arctic Climate Impact Assessment*, Cambridge University Press, Cambridge, UK

ACIA (2005) *Arctic Climate Impact Assessment: Scientific Report*, Cambridge University Press, Cambridge, UK

Adger, W. N. (2000) 'Social and ecological resilience: Are they related?', *Progress in Human Geography*, vol 24, no 3, pp347–364

Adger, W. N. (2001) 'Scales of governance and environmental justice for adaptation and mitigation to climate change', *Journal of International Development*, vol 13, pp921–931

Adger, W. N. (2006) 'Vulnerability', *Global Environmental Change*, vol 16, pp268–281

Adger, W. N. and P. M. Kelly (2001) 'Social vulnerability and resilience', in W. N. Adger, P. M. Kelly and Nguyen Huu Ninh (eds) *Living with Environmental Change: Social Vulnerability and Resilience in Vietnam*, Routledge, New York, pp19–34

Adger, W. N., K. Brown and E. L. Tompkins (2004a) 'Why do resource managers make links to stakeholders at other scales?', Tyndall Centre Working Paper No 65. Tyndall Centre for Climate Change Research, Manchester, UK

Adger, W. N., N. Brooks, G. Bentham, M. Agnew and S. Eriksen (2004b) 'New indicators of vulnerability and adaptive capacity', Technical Report 7, Tyndall Centre for Climate Change Research, Manchester, UK

Adger, W. N., N. W. Agrell and E. L. Tompkins (2005a) 'Successful adaptation to climate change across scales', *Global Environmental Change*, vol 15, pp77–86

Adger, W. N., K. Brown and E. L. Tompkins (2005b) 'The political economy of cross-scale networks in resource co-management', *Ecology and Society*, vol 10, no 2, p9, available online at www.ecologyandsociety.org/vol10/iss2/art9/, accessed February 2008

AHDR (Arctic Human Development Report) (2004) *Arctic Human Development Report*, Stefansson Arctic Institute, Akureyri, Iceland

Arts, B. (2003) 'Non-state actors in global governance: A power analysis', paper presented at the 2003 ECPR Joint Sessions, 'Workshop 11: The governance of global issues – effectiveness, accountability and constitutionalization', Edinburgh, 28 March–2 April 2003

Baerenholdt, J. O. (1996) 'The Barents Sea fisheries – New division of labour, regionalization and regionalist politicies', in J. Käkönen (ed) *Dreaming of the Barents Region: Interpreting Cooperation in the Euro-Arctic Rim*, Tampere Research Institute Research Report No 73, University of Tampere, Tampere, Finland

Bakker, K., L. del Moral, T. Downing, C. Giansante, A. Garrido, E. Iglesias, B. Pedregal and P. Riesco (1999) 'Societal and institutional responses to climate change and climatic hazards: Managing changing flood and drought risk', SIRCH (Societal and Institutional Responses to Climate Change and Climatic Hazards: Managing Changing Flood and Drought Risk) Working Paper 3, Oxford University, Oxford, UK

Berkes, F. and D. Jolly (2001) 'Adapting to climate change: Social-ecological resilience in a Canadian western Arctic community', *Conservation Ecology*, vol 5, no 2, p18, available online at www.consecol.org/vol5/iss2/art18/, accessed February 2008

Berkhout, F., J. Hertin and D. M. Gann (2004) 'Learning to adapt: Organizational adaptation to climate change impacts', Tyndall Centre Working Paper No 47, Tyndall Centre for Climate Change Research, Manchester, UK

Berkhout, F., J. Hertin and D. M. Gann (2006) 'Learning to adapt: Organizational adaptation to climate change impacts', *Climatic Change*, vol 78, pp135–156

Boland, P. (1999) 'Contested multi-level governance: Merseyside and the European Structural Funds', *European Planning Studies*, vol 7, no 5, pp647–664

Boreal Forests of the World (2003a) 'Finland – Forests and forestry', www.borealforest.org/world/world_finland.htm, accessed February 2008

Boreal Forests of the World (2003b) 'Sweden – Forests and forestry', www.borealforest.org/world/world_sweden.htm, accessed February 2008

Boreal Forest (2005) 'Boreal forests of the world', www.borealforest.org/world/world_finland.htm, accessed February 2008

Bostedt, G., P. J. Parks and M. Boman (2001) 'Integrating forestry and reindeer husbandry in northern Sweden: A discrete time application', paper submitted for presentation at the 2001 Annual Conference of the European Association of Environmental and Resource Economists (EAERE), Southampton, UK

Bostrom, A., M. G. Morgan, B. Fischhoff and D. Read (1994) 'What do people know about global climate change? 1. Mental models', *Risk Analysis*, vol 14, no 6, pp959–970

Brenner, N. 1999 'Beyond state-centrism? Space, territoriality and geographic scale in globalization studies', *Theory and Society*, vol 28, pp39–78

Brooks, N., W. N. Adger and P. M. Kelly (2005) 'The determinants of vulnerability and adaptive capacity at the national level and the implications for adaptation', *Global Environmental Change*, vol 15, pp151–163

Brooks, N. (2003) 'Vulnerability, risk and adaptation: A conceptual framework', Tyndall Centre Working Paper No 28, Tyndall Centre for Climate Change Research, Manchester, UK

Budd, L. (1998) 'Territorial competition and globalization: Scylla and Charybdis of European cities', *Urban Studies*, vol 35, no 4, pp663–685

Burgess, P. (1996) 'Deatnu: Southern habits in a northern river – Fragmentation of a river system in northern Fennoscandia', Arctic Centre Reports 17, Arctic Centre University of Lapland, Rovaniemi, Finland

Burton, I., S. Huq, B. Lim, O. Pilifosova and E. L. Schipper (2002) 'From impacts assessment to adaptation priorities: The shaping of adaptation policy', *Climate Policy*, vol 2, nos 2–3, pp145–159

Cannon, T., J. Twigg and J. Rowell (2003) 'Social vulnerability, sustainable livelihood and disasters', report to DFID Conflict and Humanitarian Assistance Department and Sustainable Livelihoods Support Office, University of Greenwich, Natural Resources Institute, London

Cash, D. W. and S. Moser (2000) 'Linking global and local scales: Designing dynamic assessment and management processes', *Global Environmental Change*, vol 10, no 2, pp109–120

Cashore, B., G. Auld and D. Newsom (2004) *Governing Through Markets: Forest Certification and the Emergence of Non-State Authority*, Yale University Press, New Haven, CT

Cerny, P. (1997) 'Paradoxes of the competition state: The dynamics of political globalization', *Government and Opposition*, vol 32, no 2, pp251–274

Chambers, R. (1989) 'Editorial introduction: Vulnerability, coping and policy', *IDS Bulletin*, vol 21, pp1–7

Chinvanno, S., S. Souvannalath, B. Lersupavithnapa, V. Kerdsuk and N. Thuan (2007) 'Strategies for managing climate risks in the Lower Mekong River Basin: A place-based approach', in N. Leary, J. Adejuwen, V. Barros, I. Burton, J. Kulkarni and R. Laseo (eds) *Climate Change and Adaptation*, Earthscan, London

Cohen, S. J. (1996) 'Integrated regional assessment of global climatic change: Lessons from the Mackenzie Basin Impact Study (MBIS)', *Global and Planetary Change*, vol 11, pp179–185

Cohen, S. J. (ed) (1997) *Mackenzie Basin Impact Study Final Report*, Environment Canada, Downsview, Ontario, Canada

Conley, T. (2002) 'The state of globalization and the globalization of the state', *Australian Journal of International Affairs*, vol 56, no 3, pp447–471

Cutter, S.L. (2001) 'Vulnerability Article 3: A research agenda for vulnerability science and environmental hazards', *Newsletter of the International Human Dimensions Programme on Global Environmental Change*, no 2

Cutter, S. L., B. J. Boruff and W. L. Shirley (2003) 'Social vulnerability to environmental hazards', *Social Science Quarterly*, vol 84, no 2, pp242–261

Dow, K., R. E. O'Connor, B. Yarnal, G. J. Carbone and C. L. Jocoy (2007) 'Why worry? Community water systems managers' perceptions of climate vulnerability', *Global Environmental Change*, vol 17, pp228–237

Dürrenberger, G., H. Kastenholz and J. Behringer (1999) 'Integrated assessment focus groups: Bridging the gap between science and policy?', *Science and Public Policy*, vol 26, no 5, October, pp341–349

Eagly, A. H. and P. Kulesa (1997) 'Attitudes, attitude structure and resistance to change. Implications for persuasion on environmental issues', in M. H. Bazerman et al (eds) *Environment, Ethics and Behaviour: The Psychology of Environmental Valuation and Degradation*, The New Lexington Press, San Franscisco, CA

Eakin, H. and M. C. Lemos (2006) 'Adaptation and the state: Latin America and the challenge of capacity-building under globalization', *Global Environmental Change*, vol 16, pp7–18

Easterling, W. E. (1997) 'Why regional studies are needed in the development of full-scale integrated assessment modelling of global change processes', *Global Environmental Change*, vol 7, no 4, pp337–356

Fankhauser, S., J. B. Smith and R. S. J. Tol (1999) 'Weathering climate change: Some simple rules to guide adaptation decisions', *Ecological Economics*, vol 30, pp67–78

Finland (1995a) 'Executive summary of the National Communication of Finland submitted under Articles 4 and 12 of the United Nations Framework Convention on Climate Change, 25 July 1995', FCCC/NC/8, Ministry of the Environment, Helsinki, http://unfccc.int/cop5/resource/docs/nc/fin01.pdf, accessed February 2008

Finland (1995b) 'Finland's National Report under the United Nations Framework Convention on Climate Change', Ministry of the Environment, Helsinki, http://unfccc.int/cop3/fccc/natcom/natc/finnc1.pdf, accessed February 2008

Finnbarents (2003) 'Finnbarents databank on economy and environment in the Barents region: Nordic parts of the Barents region', http://finnbarents.urova.fi/barentsinfo/economic/05/yla_osa5.html, accessed 24 February 2003

Finnemore, M. and K. Sikkink (1998) 'International norm dynamics and political change', *International Organisation*, vol 52, no 4, pp887–917

Finnish Forest Research Institute (2002) *Metsätilastollinen Vuosikirja 2002 [Finnish Statistical Yearbook of Forestry]*, Finnish Forest Research Institute, Vantaa, Finland, www.metla.fi/julkaisut/metsatilastollinenvsk/index-en.htm, accessed February 2008

Finnish Forest Research Institute (2006) *Finnish Statistical Yearbook of Forestry*, Finnish Forest Research Institute, Helsinki

Finnmark County Administration (2000) 'Finnmark County in the Barents Euro-Arctic Region', available at www.finnmark-f.kommune.no/, accessed 24 February 2003

FINSKEN (2003) 'Summary table of the scenario results', www.ymparisto.fi/eng/research/projects/finsken/ajankoht.htm, accessed 9 April 2003

Folke, C., L. Pritchard Jr., F. Berkes, J. Colding and U. Svedin (1998) 'IHDP Working Paper No 2', International Human Dimensions Programme on Global Environmental Change (IHDP), Bonn

Ford, J. D. and B. Smit (2004) 'A framework for assessing the vulnerability of communities in the Canadian Arctic to risks associated with climate change', *Arctic*, vol 57, no 4, pp389–400

Forsius, J., P. Kuivanlainen and P. Mäkinen (1996) 'Effects of climate change on the production and consumption of electricity in Finland', in Jaana Roos (ed) *The Finnish Research Programme on Climate Change: Final Report*, Publications of the Academy of Finland 4/96, Edita, Helsinki

Fraser, E. D. G. (2007) 'Travelling in antique lands: Using past famines to develop an adaptability/resilience framework to identify food systems vulnerable to climate change', *Climatic Change*, vol 83, pp495–514

Füssel, H-M. (2007) 'Vulnerability: A generally applicable framework for climate change research', *Global Environmental Change*, vol 17, pp155–167

Füssel, H-M. and R. J. T. Klein (2006) 'Climate change vulnerability assessments: An evolution of conceptual thinking', *Climatic Change*, vol 75, pp301–329

Gateway to Sweden (2005) 'Swedish industry', www.sweden.se/templates/cs/BasicFactsheet_6900.aspx, accessed 2 September 2005

Gissendanner, S. (2003) 'Methodology problems in urban governance studies', *Environment and Planning C: Government and Policy*, vol 21, pp663–685

Gjøsaeter, H. (1995) 'Pelagic fish and the ecological impact of the modern fishing industry in the Barents Sea', *Arctic*, vol 48, no 3, pp267–278

Goodwin, P. (1998) '"Hired hands" or "local voice": Understandings and experience of local participation in conservation', *Transactions of the Institute of British Geographers*, New Series, vol 23, no 4, pp481–499

Gregory, R., B. Fischhoff, S. Thornec and G. Buttec (2003) 'A multi-channel stakeholder consultation process for transmission deregulation', *Energy Policy*, vol 31, pp1291–1299

Guillén, M. F. (2001) 'Is globalization civilizing, destructive or feeble? A critique of five key debates in the social science literature', *Annual Review of Sociology*, vol 27, pp235–260

Guisan, A., J. I. Holten, R. Spichiger and L. Tessier (eds) (1995) *Potential Ecological Impacts of Climate Change in the Alps and Fennoscandian Mountains. An Annex to the Intergovernmental Panel on Climate Change (IPCC) Second Assessment Report, Working Group II-C (Impacts of Climate Change on Mountain Regions)*, Conservatoire et Jardin Botaniques de Genève, Geneva

Hallenstvedt, A. (1993) 'Ressurskontroll i norsk fiskeriforvaltning', in Nordic Council of Ministers (1993) *Ressursforvaltning og Kontroll – Kontrollpolitikk og Virkemidler i Fiskeriene i de Nordiske Land og i EF*, Nordiske Seminar og Arbejdsrapporter 1993, 583, Nordic Council of Ministers, Copenhagen

Hannelius, S. and K. Kuusela (1995) *Finland the Country of Evergreen Forest*, Forssan Kirjapaino oy, Helsinki

Heal, O. W., T. V. Callaghan, J. H. C. Cornelissen, C. Körner and S. E. Lee, (1998) 'Global change in Europe's cold regions', Ecosystems Research Report No 27, European Commission, Luxembourg

Held, D. and A. McGrew (2002) 'Introduction', in D. Held and A. McGrew (eds) *Governing Globalization: Power, Authority and Global Governance*, Polity Press, Cambridge, UK

Held, D., A. McGrew, D. Goldblatt and J. Perraton (1999) *Global Transformations: Politics, Economics and Culture*, Polity Press, Cambridge, UK

Henttonen, H. (1995) 'Climate change and the ecology of alpine mammals', in A. Guisan, J. I. Holten, R. Spichiger and L. Tessier (eds) *Potential Ecological Impacts of Climate Change in the Alps and Fennoscandian Mountains. An Annex to the Intergovernmental Panel on Climate Change (IPCC) Second Assessment Report, Working Group II-C (Impacts of Climate Change on Mountain Regions)*, Conservatoire et Jardin Botaniques de Genève, Geneva

Hoel, A. H. (1994) 'The Barents Sea: Fisheries resources for Europe and Russia', in O. S. Stokke and O. Tunander (eds) *The Barents Region: Cooperation in Arctic Europe*, Sage, London and Thousand Oaks, CA

Høgda, K. A., S. R. Karlsen and I. Solheim (2001) 'Climatic change impact on growing season in Fennoscandia studied by a time series of NOAA AVHRR NDVI data', *Proceedings of the International Geoscience and Remote Sensing Symposium IGARSS*, 9–13 July 2001, Sydney, Australia (ISBN 978-0-7803-7536-9)

Hollis, M. and S. Smith (1991) *Explaining and Understanding International Relations*, Oxford University Press, Oxford, UK

Holman, I. P., M. D. A. Rounsewell, S. Shackley, P. A. Harrison, R. J. Nicholls, P. M. Berry and E. Audsley (2005) 'A regional, multi-sectoral and integrated assessment of the impacts of climate and socioeconomic change in the UK', *Climatic Change*, vol 71, pp9–41

Holten, J. I. (1995) 'Effects of climate change on plant diversity and distribution', in A. Guisan, J. I. Holten, R. Spichiger and L. Tessier (eds) *Potential Ecological Impacts of Climate Change in the Alps and Fennoscandian Mountains. An Annex to the Intergovernmental Panel on Climate Change (IPCC) Second Assessment Report, Working Group II-C (Impacts of Climate Change on Mountain Regions)*, Conservatoire et Jardin Botaniques de Genève, Geneva

Hooghe, L. and G. Marks (2003) 'Unravelling the central state, but how? Types of multi-level governance', *American Political Science Review*, vol 97, no 2, pp233–243

Huq, S., A. Kokorin, M. Guedelha and A. Narayan Achanta (1999) 'Finland Report on the in-depth review of the Second National Communication of Finland 20 May 1999', FCCC/IDR.2/FIN, http://unfccc.int/cop5/resource/docs/idr/fin02.pdf, accessed February 2008

IPCC (1998) *The Regional Impacts of Climate Change: An Assessment of Vulnerability*, Cambridge University Press, Cambridge, UK

IPCC (J. J. McCarthy, O. F. Canziani, N. A. Leary, D. J. Dokken and K. S. White (eds)) (2001) *Climate Change 2001: Mitigation. Contribution of Working Group III to the Third Assessment Report of the Intergovernmental Panel on Climate Change*, Cambridge University Press, Cambridge, UK

Jepma, C. J. and M. Munasinghe (1998) *Climate Change Policy: Facts, Issues and Analyses*, Cambridge University Press, Cambridge, UK

Jernsletten, J.-L. and H. Beach (2006) 'The challenges and dilemmas of concession reindeer management in Sweden', in B. C. Forbes, M. Bölter, L. Muller-Wille, J. Hukkinen, F. Müller, N. Gunslay and Y. Konstantinov (eds) *Reindeer Management in Northernmost Europe: Linking Practical and Scientific Knowledge in Social-Ecological Systems*, Springer-Verlag, Berlin

Jones, S. A., B. Fischhoff and D. Lach (1999) 'Evaluating the science–policy interface for climate change research', *Climatic Change*, vol 43, pp581–599

Jürgens, I. (2002) 'Science-stakeholder dialogue and climate change: Towards a participatory notion of communication', paper presented at the 2002 Berlin Conference on the Human Dimensions of Global Environmental Change, 'Knowledge for sustainability transition: The challenge for social science', Berlin, 6–7 December

Kates, R. W., J. H. Ausubel and M. Berberian (eds) (1985) *Climate Impact Assessment: Studies of the Impact of Climate and Society*, John Wiley and Sons, Chichester, UK

Kelly, P. M. and W. N. Adger (2000) 'Theory and practice in assessing vulnerability to climate change and facilitating adaptation', *Climatic Change*, vol 47, pp325–352

Kempton, W. (1991) 'Lay perspectives on global climate change', *Global Environmental Change*, vol 1, pp183–208

Kennedy, P. (2007) 'Global transformations but local, "bubble" lives: Taking a reality check on some globalization concepts', *Globalizations*, vol 4, no 2, pp267–282

Keohane, R. O. and J. S. Nye, Jr. (2000) 'Governance in a globalizing world', in J. S. Nye and J. D. Donahue (eds) *Governance in a Globalizing World*, Brookings Institution Press, Washington, DC

Keskitalo, E. C. H. (2004a) 'A framework for multi-level stakeholder studies in response to global change', *Local Environment*, vol 9, no 5, pp425–435

Keskitalo, E. C. H. (2004b) *Negotiating the Arctic: The Construction of an International Region*, Routledge, New York and London

Klokov, K. and J.-L. L. Jernsletten (2002) *Sustainable Reindeer Husbandry: A Report for the Arctic Council 2000–2002*, Centre for Saami Studies, University of Tromsø, Tromsø, Norway

Klungsøyr, J., R. Saetre, L. Føyn and H. Loeng (1995) 'Man's impact on the Barents Sea', *Arctic*, vol 48, no 3, pp279–296

Knapp, G., P. Livingston and A. Tyler. (1999) 'Human effects of climate-related changes in Alaska commercial fisheries', in G. Weller and P. A. Anderson (eds) *Assessing the Consequences of Climate Change for Alaska and the Bering Sea Region. Proceedings of a Workshop at the University of Alaska Fairbanks, 29–30 October 1998*, Center for Global Change and Arctic System Research, University of Alaska, Fairbanks, AK

Kok, K., M. Patel, D. S. Rothman and S. Greeuw (2002) 'First series of target area workshops. October–November 2002. Process and methodology', *MedAction Deliverable #6,* Working Paper I03-E002, International Centre for Intergrative Studies ICIS, Maastricht, The Netherlands

Krogh, L. (1995) 'Information technology in the Norwegian fishing industry', in S. Johansen (ed) *Nordiske fiskersamfund i fremtiden – Vol. 1: Fiskeri og Fiskersamfund*, TemaNord, No 585, Nordic Council of Ministers, Copenhagen

Kumpula, J. (2001) 'Productivity of the semi-domesticated reindeer (*Rangifer t. tarandus* L.) stock and carrying capacity of pastures in Finland during 1960s–1990s', PhD dissertation, Department of Biology, University of Oulu, Reindeer Research Station and Finnish Game and Fisheries Research Institute, Finland, http://herkules.oulu.fi/isbn9514265556/html/index.html, accessed February 2008

Kuoppamäki, P. (1996a) *Impacts of Climate Change from a Small Nordic Open Economy Perspective*, The Research Institute of the Finnish Economy, Helsinki

Kuoppamäki, P. (1996b) 'The impacts of climate change on the Finnish economy', in Jaana Roos (ed) *The Finnish Research Programme on Climate Change: Final Report*, Publications of the Academy of Finland 4/96, Edita, Helsinki

Lafferty, W. M. and J. Meadowcroft (1996) 'Democracy and the environment: Congruence and conflict – Preliminary reflections', in W. M. Lafferty and J. Meadowcroft (eds) *Democracy and the Environment: Problems and Prospects*, Edward Elgar, Cheltenham, UK

Lange, M. A (1999) 'Integrated regional impact studies in the European north: Basic issues, methodologies and regional climate modelling (IRISEN): A summary', in M. A. Lange (ed) *IRISEN Integrated Regional Impact Studies in the European North. Proceedings of an Advanced Study Course at Abisko Research Station, Sweden, 4–16 July 1999*, Institute for Geophysics, University of Münster, Münster, Germany

Lange, M. (2001) 'The Barents Sea Impact Study (BASIS): Methodology and first results', paper submitted to *Proceedings of the ELOISE Open Science Meeting*, Rende, Italy

Länsarbetsnämnden and Samernas Riksförbund (1996) *Inventering av den Samiska sysselsättningsstrukturen i Norrbottens län [Inventory of the Saami Occupational Structure in Norrbotten county]*, Samernas Riksförbund, Umeå, Sweden

Layton, I. and A. Pashkevitch (2000) 'Forest-sector case studies in the Barents region: Comparisons between the Northern Dvina and the Pite river basins', draft working report, Department of Social and Economic Geography, Umeå University, Umeå, Sweden

Lazo, J. K., J. C. Kinnell and A. Fisher (2000) 'Expert and layperson perceptions of ecosystem risk', *Risk Analysis*, vol 20, no 2

Leary, N., J. Adejuwon, V. Barros, P. Batimaa, B. Biagini, I. Burton, S. Chinvanno, R. Cruz, D. Dabi, A. de Comarmond, B. Dougherty, P. Dube, A. Githeko, A. A. Hadid, M. Hellmuth, R. Kangalawe, J. Kulkarni, M. Kumar, R. Lasco, M. Mataki, M. Medany, M. Mohsen, G. Nagy, M. Njie, J. Nkomo, A. Nyong, B. Osman-Elasha, E. Sanjak, R. Seiler, M. Taylor, M. Travasso, G. von Maltitz, S. Wandiga and M. Wehbe (2007) 'A stitch in time: General lessons from specific cases', in N. Leary, J. Adejuwen, V. Barros, I. Burton, J. Kulkarni and R. Laseo (eds) *Climate Change and Adaptation*, Earthscan, London

Lebel, L., P. Garden and M. Imamura (2005) 'The politics of scale, position and place in the governance of water resources in the Mekong region', *Ecology and Society*, vol 10, no 2, p18, available online at www.ecologyandsociety.org/vol10/iss2/art18/, accessed February 2008

Lee, S. E. (1999) 'Vegetation modelling with specific emphasis on the Arctic and sub-Arctic environment', in M. A. Lange (ed) (2001) *IRISEN Integrated Regional Impact Studies in the European North. Proceedings of an Advanced Study Course at Abisko Research Station, Sweden, 4–16 July 1999*, Institute for Geophysics, University of Münster, Münster, Germany

Lim, B. and I. Burton (2001) *An Adaptation Policy Framework: Capacity Building for Stage II Adaptation*, UNDP-GEF Project, National Communications Support Programme, Global Environment Facility

Lindgren, U., Ö. Pettersson, B. Jansson and H. Nilsagård (2000) *Skogsbruket i den lokala ekonomin [Forestry in the Local Economy]*, Report No 4, Swedish Forest Agency, Jönköping, Sweden

Liverman, D. M. (1990) 'Drought impacts in Mexico: Climate, agriculture, technology and land tenure in Sonora and Puebla', *Annals of the Association of American Geographers*, vol 80, pp49–72

Löfgren, H. and M. Benner (2003) 'Biotechnology and governance in Australia and Sweden: Path dependency or institutional convergence', *Australian Journal of Political Science*, vol 38, no 1, pp25–53

Luers, A. L. (2005) 'The surface of vulnerability: An analytical framework for examining environmental change', *Global Environmental Change*, vol 15, pp214–223

MacIver, D. C. and E. Wheaton (2005) 'Tomorrow's forests: Adapting to climate change', *Climatic Change*, vol 70, pp273–282

Mahler, V. A. (2004) 'Economic globalization, domestic politics and income inequality in the developed countries: A cross-national study', *Comparative Political Studies*, vol 37, no 9, pp1025–1053

Mariussen, Å. and K. Heen (1999) 'Dependence, uncertainty and climate change: The case of fishing', in M. A. Lange, B. Bartling and K. Grosfeld (eds) *Global Changes and the Barents Sea Region. Proceedings of the First International BASIS Research Conference, St. Petersburg, Russia, 22–25 February 1998*, Institute for Geophysics, University of Münster, Münster, Germany, pp335–338, available at http://earth.uni-muenster.de/BASIS/BASISP/MarHeen.html, accessed 5 May 2000

Marsden, T. K. (1997) 'Reshaping environments: Agriculture and water interactions and the creation of vulnerability', *Transactions of the Institute of British Geographers*, New Series, vol 22, no 3, pp321–337

May, T. (2001) *Social Research: Issues, Methods and Process*, Open University Press, Buckingham, UK

McDaniels, T. L., H. Dowlatabadi and S. Stevens (2005) 'Multiple scales and regulatory gaps in environmental change: The case of salmon aquaculture', *Global Environmental Change*, vol 15, pp9–21

Mendelsohn, R. (2006) 'The role of markets and governments in helping society adapt to a changing climate', *Climatic Change*, vol 78, pp203–215

Moench, M. H. (2004) 'Adaptive capacity and livelihood resilience in water scarce areas: Research results from South Asia and implications for the Middle East', paper presented at 'Water for life in the Middle East', second Israeli–Palestinian international conference, Antalya, Turkey, 10–14 October 2004

Moench, M. and A. Dixit (eds) (2004) *Adaptive Capacity and Livelihood Resilience: Adaptive Strategies for Responding to Floods and Draught in South Asia*, Institute for Social and Environmental Transition, Boulder, CO

Molau, U. (1995) 'Climate change, plant reproductive ecology and population dynamics', in A. Guisan, J. I. Holten, R. Spichiger and L. Tessier (eds) *Potential Ecological Impacts of Climate Change in the Alps and Fennoscandian Mountains. An Annex to the Intergovernmental Panel on Climate Change (IPCC) Second Assessment Report, Working Group II-C (Impacts of Climate Change on Mountain Regions)*, Conservatoire et Jardin Botaniques de Genève, Geneva

Mostert, E. (2002) 'The challenge of public participation', *Water Policy*, vol 5, no 2, pp179–197

Myrstad, B. (2000) 'The Norwegian fishing and aquaculture industry', http://odin.dep.no/odin/engelsk/norway/economy/032001-990373/index-dok000-b-n-a.html, accessed 2 September 2005

Naess, L. O., G. Bang, S. Eriksen and J. Vevatne (2005) 'Institutional adaptation to climate change: Flood responses at the municipal level in Norway', *Global Environmental Change*, vol 15, pp125–138

Naess, L. O., I. Thorsen Norland, W. M. Lafferty and C. Aall (2006) 'Data and processes linking vulnerability assessment to adaptation decision-making on climate change in Norway', *Global Environmental Change*, vol 16, pp221–233

Nagy, G. J., M. Bidegain, R. M. Caffera, W. Norbis, A. Ponce, V. Pshennikov and D. N. Severov (2007) 'Fishing strategies for managing climate variability and change in the estuarine front of the Rio de la Plata', in N. Leary, J. Adejuwen, V. Barros, I. Burton, J. Kulkarni and R. Laseo (eds) *Climate Change and Adaptation*, Earthscan, London

Norén, R. (2004) 'Dismissal of employees in Swedish manufacturing', *Journal of Policy Modeling*, vol 26, pp151–164

Nordic Council of Ministers (2002) *Have a 'Good Participation': Recommendations on Public Participation in Forestry Based on Literature Review and Nordic Experiences*, TemaNord No. 515, Nordic Council of Ministers, Copenhagen

Norrbotten County Administration (2002) *Sveaskogsprojektet – Inventering av skyddsvärda skogar på Sveaskogs marker i Norrbottens län [The Sveaskog project – Inventory of Forest with Protection Values owned by Sveaskog in Norrbotten County]*, Rapport 7:2002, www.bd.lst.se/livsmiljo/naturvard/0702.htm, accessed February 2008

Norway (1994a) 'Executive Summary of the National Communication of Norway submitted under Articles 4 and 12 of the United Nations Framework Convention on Climate Change A/AC.237/NC/11, 26 October 1994', Norwegian Pollution Control Authority SFT, Olso, http://unfccc.int/cop5/resource/docs/nc/nor01.pdf, accessed February 2008

Norway (1994b) 'Norway's national communication under the Framework Convention on Climate Change – September 1994', http://unfccc.int/cop3/fccc/natcom/natc/nornc1.pdf, accessed February 2008

Nuttall, M. (2005) 'Hunting, fishing and gathering: Indigenous peoples and renewable resource use in the Arctic', in ACIA (Arctic Climate Impact Assessment) (ed) *Arctic Climate Impact Assessment: Scientific Report*, Cambridge University Press, Cambridge, UK, pp649–690

Nyong, A., D. Dabi, A. Adeptu, A. Berthe and C. Ihemegbulem (2007) 'Vulnerability in the Sahelian zone of northern Nigeria: A household-level assessment', in N. Leary, C. Conde, J. Kulkarni, A. Nyong and J. Pulhin (eds) *Climate Change and Vulnerability*, Earthscan, London

O'Brien, K. L. and R. M. Leichenko (2000) 'Double exposure: Assessing the impacts of climate change within the context of economic globalization', *Global Environmental Change*, vol 10, pp221–232

O'Brien, K., L. Sygna and J. E. Haugen (2004a) 'Vulnerable or resilient? A multi-scale assessment of climate impacts and vulnerability in Norway', *Climatic Change*, vol 64, pp193–225

O'Brien, K., R. Leichenko, U. Kelkar, H. Venema, G. Aandahl, H. Tompkins, A. Javed, S. Bhadwal, S. Barg, L. Nygaard and J. West (2004b) 'Mapping vulnerability to multiple stressors: Climate change and globalization in India', *Global Environmental Change*, vol 14, pp303–313

Osman-Elasha, B. and E. A. Sanjak (2007) 'Livelihoods and drought in Sudan', in N. Leary, C. Conde, J. Kulkarni, A. Nyong and J. Pulhin (eds) *Climate Change and Vulnerability*, Earthscan, London

Ostrom, E. (2001) 'Vulnerability and polycentric governance systems', *IHDP Update*, Newsletter of the International Human Dimensions Programme on Global Environmental Change, No 3

Ouedraogo, A., F. Senhaji, M. Thorstensson and T. Dallman (1999) 'Report on the in-depth review of the second national communication of Norway, 1 July 1999', FCCC/IDR.2/NOR, http://unfccc.int/resource/docs/idr/nor02.pdf, accessed February 2008

Paavola, J. and W. N. Adger (2002) 'Knowledge or/and participation for sustainability? Science, justice and environmental governance in adaptation to climate change', paper presented at the 2002 Berlin Conference on the Human Dimensions of Global Environmental Change Knowledge for Sustainability Transition: The Challenge for Social Science, Berlin, 6–7 December 2002

Patt, A., R. J. T. Klein and A. de la Vega-Leinert (2005) 'Taking the uncertainty in climate-change vulnerability assessment seriously', *Comptes Rendus Geosciences*, vol 337, pp411–424

Pattberg, P. (2005) 'The institutionalization of private governance: How business and nonprofit organizations agree on transnational rules', *Governance*, vol 18, no 4, pp589–610

Pedras, M. (2002) 'From local consciousness to global change: Asserting power at the local scale', *International Journal of Urban and Regional Research*, vol, 26, no 4, pp823–833

Pelling, M., C. High, J. Dearing and D. Smith (2007) 'Shadow spaces for social learning: A relational understanding of adaptive capacity to climate change within organizations', *Environment and Planning A*, forthcoming (advance online publication, doi:10.1068/a39148)

Pettersson, Ö. (2002) *Socio-Economic Dynamics in Sparse Regional Structures*, PhD dissertation, GERUM Reports in Social and Economic Geography No 2, Department of Social and Economic Geography, Umeå University, Umeå, Sweden

Petts, J., T. Horlick-Jones and G. Murdock (2001) 'Social amplification of risk: The media and the public', Contract Research Report 329/2001, Health and Safety Executive, Norwich, UK

Polsky, C., D. Schröter, A. Patt, S. Gaffin, M. Long Martello, R. Neff, A. Pulsipher and H. Selin (2003) 'Assessing vulnerabilities to the effects of global change: An eight-step approach', Research and Assessment Systems for Sustainability Program Discussion Paper 2003–05, Environment and Natural Resources Program, Belfer Center for Science and International Affairs, Kennedy School of Government, Harvard University, Cambridge, MA

Rankin, K. N. (2003) 'Anthropologies and geographies of globalization', *Progress in Human Geography*, vol 27, no 6, pp708–734

Rayner, S. and E. L. Malone (eds) (1998) *Human Choice and Climate Change. Volume Three: The Tools for Policy Analysis*, Battelle Press, Columbus, OH

Riissanen, N. and J. Härkönen (eds) (2000) *Lapin Metsäohjelma [Lapland Forest Programme] 2001–2005*, Metsäkeskus, Rovaniemi, Finland

Rikkinen, K. (1992) *A Geography of Finland*, University of Helsinki, Lahti Research and Training Centre, Lahti, Finland

Robertson, R. (1992) *Globalization: Social Theory and Global Culture*, Sage Publications, London

Rotmans, J. (2006) 'Tools for integrated sustainability assessment: A two-track approach', *Integrated Assessment Journal*, vol 6, no 4, pp35–57

Roudometof, V. (2003) 'Glocalization, space and modernity', *The European Legacy*, vol 8, no 1, pp37–60

Saelthun, N. R. (1995) 'Climate change impact on hydrology and cryology', in A. Guisan, J. I. Holten, R. Spichiger and L. Tessier (eds) *Potential Ecological Impacts of Climate Change in the Alps and Fennoscandian Mountains. An Annex to the Intergovernmental Panel on Climate Change (IPCC) Second Assessment Report, Working Group II-C (Impacts of Climate Change on Mountain Regions)*, Conservatoire et Jardin Botaniques de Genève, Geneva

Sandström, C. and E. I. Falleth (eds) (2008) *Omstridda naturresurser. Trender och utmaningar i nordisk naturresursförvaltning [Conflicted Environmental Resources: Trends and Challenges in Nordic Environmental Resource Management]*, Boréa Förlag, Umeå, Sweden

Schiller, A., C. T. Hunsaker, M. A. Kane, A. K. Wolfe, V. H. Dale, G. W. Suter, C. S. Russell, G. Pion, M. H. Jensen and V. C. Konar (2001) 'Communicating ecological indicators to decision-makers and the public', *Conservation Ecology*, vol 5, no 1, available online at www.ecologyandsociety.org/vol5/iss1/art19/, accessed February 2008

Schneider, S. H. (1997) 'Integrated assessment modelling of global climate change: Transparent rational tool for policymaking or opaque screen hiding value-laden assumptions?', *Environmental Modeling and Assessment*, vol 2, pp229–249

Schneider, S. and J. Sarukhan (2001) 'Overview of impacts, adaptation and vulnerability to climate change', in J. J. McCarthy, O. F. Canziani, N. A. Leary, D. J. Dokken and K. S. White (eds) *Climate Change 2001: Impacts, Adaptation, and Vulnerability*, contribution of Working Group II to the Third Assessment Report of the Intergovernmental Panel on Climate Change, published for the Intergovernmental Panel on Climate Change, Cambridge University Press, Cambridge, UK

Schröter, D., C. Polsky and A.G. Patt (2005) 'Assessing vulnerabilities to the effects of global change: An eight step approach', *Mitigation and Adaptation Strategies for Global Change*, vol 10, no 4, pp573–595

Scruggs, L. and P. Lange (2002) 'Where have all the members gone? Globalization, institutions and union density', *The Journal of Politics*, vol 64, no 1, pp126–153

Selnes, M. (1995) 'Fish resources and management from a regional perspective: Outlining a regional enterprise quota system', in S. Johansen (ed) *Nordiske fiskersamfund i fremtiden – Vol 1: Fiskeri og fiskersamfund*, TemaNord No 585, Nordic Council of Ministers, Copenhagen

Seppänen, S. (1995) *The Barents Region: An Emerging Market*, Statistics Finland, Helsinki

Shackley, S. and C. Gough (2002) 'The use of integrated assessment: An institutional analysis perspective', Tyndall Centre Working Paper No 14, Tyndall Centre for Climate Change Research, Manchester, UK

Shelley, B. (2000) 'Political globalization and the politics of international non-governmental organizations: The case of village democracy in China', *Australian Journal of Political Science*, vol 35, no 2, pp225–238

Sites, W. (2000) 'Primitive globalization? State and locale in neoliberal global engagement', *Sociological Theory*, vol 18, no 1, pp121–144

Skre, O. (1999) 'Climate change impacts on mountain birch ecosystems', in M. A. Lange (ed) (2001) *IRISEN Integrated Regional Impact Studies in the European North. Proceedings of an Advanced Study Course at Abisko Research Station, Sweden, 4–16 July 1999*, Institute for Geophysics, University of Münster, Münster, Germany

Smit, B. and O. Pilifosova (2001) 'Adaptation to climate change in the context of sustainable development and equity', in J. J. McCarthy, O. F. Canziani, N. A. Leary, D. J. Dokken and K. S. White (eds) *Climate Change 2001: Impacts, Adaptation, and Vulnerability*, contribution of Working Group II to the Third Assessment Report of the Intergovernmental Panel on Climate Change, published for the Intergovernmental Panel on Climate Change, Cambridge University Press, Cambridge, UK

Smit, B. and M. W. Skinner (2002) 'Adaptation options in agriculture to climate change: A typology', *Mitigation and Adaptation Strategies for Global Change*, vol 7, pp85–114

Smit, B. and J. Wandel (2006) 'Adaptation, adaptive capacity and vulnerability', *Global Environmental Change*, vol 16, pp282–292

Smit, B., I. Burton, R. J. T. Klein and J. Wandel (2000) 'An anatomy of adaptation to climate change and variability', *Climatic Change*, vol 45, pp223–251

Smith, J. B., H.-J. Schellnhuber and M. M. Qader Mirza (2001) 'Vulnerability to climate change and reasons for concern: A synthesis', in J. J. McCarthy, O. F. Canziani, N. A. Leary, D. J. Dokken and K. S. White (eds) *Climate Change 2001: Impacts, Adaptation, and Vulnerability*, contribution of Working Group II to the Third Assessment Report of the Intergovernmental Panel on Climate Change, published for the Intergovernmental Panel on Climate Change, Cambridge University Press, Cambridge, UK

Smithers, J., and B. Smit (1997) 'Human adaptation to climatic variability and change', *Global Environmental Change*, vol 7, no 2, pp129–146

Snyder, F. (1999) 'Governing economic globalization: Global legal pluralism and European law', *European Law Journal*, vol 5, no 4, pp334–374

SOU (2001) *En ny rennäringspolitik – öppna samebyar och samverkan med andra markanvändare [A New Reindeer Herding Policy: Open Saami Villages and Cooperation with other Land Uses]*, Swedish Government Official Reports SOU 2001:101, Stockholm, Sweden

Stonich, S. C. and J. R. Bort (1997) 'Globalization of shrimp mariculture: The impact on social justice and environmental quality in Central America', *Society and Natural Resources*, vol 10, no 2, pp161–180

Sweden (1994) 'Sweden's national report under the United Nations Framework Convention on Climate Change', Ministry of the Environment and Natural Resources, http://unfccc.int/resource/docs/natc/swenc1.pdf, accessed February 2008

Sweden (1997) 'Sweden's second national communication on climate change', http://unfccc.int/resource/docs/natc/swenc2.pdf, accessed February 2008

Swedish National Board of Forestry (2004) *Swedish Statistical Yearbook of Forestry*, Swedish National Board of Forestry, Jönköping, Sweden

Swedish Forest Agency (2006) "Swedish Forest Agency". Online at www.skogsstyrelsen.se/, accessed February 2008

Sygna, L. and K. O'Brien (2001) *Virkninger av klimaendringer i Norge. Oppsummeringsrapport fra seminaret i Oslo, 30. og 31 oktober 2000 [Impacts of Climate Change in Norway: Summary Report from the Seminar in Oslo, 30–31 October 2000]*, CICERO Report 2001:1, Center for International Climate and Environmental Research, Oslo University, Oslo

Sygna, L., S. Eriksen, K. O'Brien and L. O. Naess (2004) 'Climate change in Norway: Analysis of economic and social impacts and adaptations', CICERO Report 2004:12, Center for International Climate and Environmental Research, Oslo University, Oslo

Sykora, L. (1994) 'Local urban restructuring as a mirror of globalization processes: Prague in the 1990s', *Urban Studies*, vol 31, no 7

Tarrow, S. (1999) 'International institutions and contentious politics: Does internationalization make agents freer – Or weaker?', paper presented to the Panel on Coping with World Transitions, American Sociological Association Annual Meeting, Chicago, IL, 6 August 1999

Tenbrunsel, A. E., K. A. Wade-Benzoni, D. M. Messick and M. H. Bazerman (1997) 'Introduction', in M. H. Bazerman, D. M. Messick, A. E. Tenbrunsel and K. A. Wade-Benzoni (eds) *Environment, Ethics and Behaviour: The Psychology of Environmental Valuation and Degradation*, New Lexington Press, San Fransisco, CA

Tengö, M. and K. Belfrage (2004) 'Local management practices for dealing with change and uncertainty: A cross-scale comparison of cases in Sweden and Tanzania', *Ecology and Society*, vol 9, no 3, available online at www.ecologyandsociety.org/vol9/iss3/art4, accessed February 2008

Thompson, A., P. Robbins, B. Sohngen, J. Arwai and T. Koontz (2006) 'Economy, politics and institutions: From adaptation to adaptive management in climate change', *Climatic Change*, vol 78, pp1–5

Timmerman, P. (1981) *Vulnerability, Resilience and the Collapse of Society*, Environmental Monograph 1, Institute for Environmental Studies, Toronto University, Toronto, Canada

Tol, R. S. J. and G. W. Yohe (2007) 'The weakest link hypothesis for adaptive capacity: An empirical test', *Global Environmental Change*, vol 17, pp218–227

Tol, R. S. J., S. Fankhauser and J. B. Smith (1998) 'The scope for adaptation to climate change: What can we learn from the impact literature?', *Global Environmental Change*, vol 8, no 2, pp109–123

Tompkins, E. L. and W. N. Adger (2003) 'Building resilience to climate change through adaptive management of natural resources', Tyndall Centre Working Paper No 27, Tyndall Centre for Climate Change Research, Manchester, UK

Tompkins, E., W. N. Adger and K. Brown (2002) 'Institutional networks for inclusive coastal management in Trinidad and Tobago', *Environment and Planning A*, vol 34, pp1095–1111

Town of Kemi (2003) Town of Kemi homepages, www.kemi.fi/english/, accessed February 2008

Turner, B. L. II, R. E. Kasperson, P. A. Matson, J. J. McCarthy, R. W. Corell, L. Christensen, N. Eckley, J. X. Kasperson, A. Luers, M. L. Martello, C. Polsky, A. Pulsipher and A. Schiller (2003) 'A framework for vulnerability analysis in sustainability science', *PNAS*, vol 100, no 14, pp8074–8079

Tykkyläinen, M. (1988) 'Periphery syndrome – A reinterpretation of regional development theory in a resource periphery', Reprint from *Fennia*, vol 166, no 2, pp295–411, Geographical Society of Finland, Helsinki

UK Climate Impacts Programme (2001) 'Socioeconomic scenarios for climate change impact assessment: A guide to their use in the UK Climate Impacts Programme', UKCIP, Oxford, accessed 9 April 2003 from www.ukcip.org.uk/ukcip.html

Van Asselt, M. B. A. and N. Rijkens-Klomp (2002) 'A look in the mirror: Reflection on participation in integrated assessment from a methodological perspective', *Global Environmental Change*, vol 12, pp167–184

Vilhjálmsson, H. And A. H. Hoel (2005) 'Fisheries and aquaculture', in ACIA (Arctic Climate Impact Assessment) (eds) *Arctic Climate Impact Assessment: Scientific Report*, Cambridge University Press, Cambridge, UK, pp691–780

Walker, B., S. Carpenter, J. Anderies, N. Abel, G. Cumming, M. Janssen, L. Lebel, J. Norberg, G. D. Peterson and R. Pritchard (2002) 'Resilience management in social-ecological systems: A working hypothesis for a participatory approach', *Conservation Ecology*, vol 6, no 1, available online at www.consecol.org/vol6/iss1/art14, accessed February 2008

Wiberg, U. (1980) 'Sektorer i samspel för glesbygdsutveckling i balans – Konsekvensanalys av tendenser och planeringsinsatser på lokal och regional arbetsmarknad' ['Sectors in cooperation for a balanced rural area development: An analysis of tendencies and planning approaches for local and regional employment'], Report B:4, Department of Geography, Umeå University, Umeå, Sweden

Wilbanks, T. J. (2002) 'Geographic scaling issues in integrated assessments of climate change', *Integrated Assessment*, vol 3, nos 2–3, pp100–114

Yin, Y., N. Clinton, B. Luo and L. Song (2007) 'Resource system vulnerability to climate stresses in the Heihe river basin of western China', in N. Leary, C. Conde, J. Kulkarni, A. Nyong and J. Pulhin (eds) *Climate Change and Vulnerability*, Earthscan, London

Young, O. R. (2002) *The Institutional Dimensions of Environmental Change: Fit, Interplay, Scale*, MIT Press, Cambridge, MA

Young, O. R., F. Berkhout, G. C. Gallopin, M. A. Janssen, E. Ostrom and S. van der Leeuw (2006) 'The globalization of socio-ecological systems: An agenda for scientific research', *Global Environmental Change*, vol 16, pp304–316

Ziervogel, G. and T. E. Downing (2004) 'Stakeholder networks: Improving seasonal climate forecasts', *Climatic Change*, vol 65, pp73–101

Index